Engineering Societies and Undergraduate Engineering Education

PROCEEDINGS OF A WORKSHOP

Steve Olson, *Rapporteur*

NATIONAL ACADEMY OF ENGINEERING

THE NATIONAL ACADEMIES PRESS
Washington, DC
www.nap.edu

THE NATIONAL ACADEMIES PRESS 500 Fifth Street, NW Washington, DC 20001

This activity was supported by grant No. EEC-1360962 from the National Science Foundation. Any opinions, findings, conclusions, or recommendations expressed in this publication do not necessarily reflect the views of any organization or agency that provided support for the project.

International Standard Book Number-13: 978-0-309-46466-6
International Standard Book Number-10: 0-309-46466-8
Digital Object Identifier: https://doi.org/10.17226/24878

Additional copies of this publication are available for sale from the National Academies Press, 500 Fifth Street, NW, Keck 360, Washington, DC 20001; (800) 624-6242 or (202) 334-3313; http://www.nap.edu.

Printed in the United States of America

Suggested citation: National Academy of Engineering. 2017. *Engineering Societies and Undergraduate Engineering Education: Proceedings of a Workshop.* Washington, DC: The National Academies Press. doi: https://doi.org/10.17226/24878.

The National Academies of

SCIENCES • ENGINEERING • MEDICINE

The **National Academy of Sciences** was established in 1863 by an Act of Congress, signed by President Lincoln, as a private, nongovernmental institution to advise the nation on issues related to science and technology. Members are elected by their peers for outstanding contributions to research. Dr. Marcia McNutt is president.

The **National Academy of Engineering** was established in 1964 under the charter of the National Academy of Sciences to bring the practices of engineering to advising the nation. Members are elected by their peers for extraordinary contributions to engineering. Dr. C. D. Mote, Jr., is president.

The **National Academy of Medicine** (formerly the Institute of Medicine) was established in 1970 under the charter of the National Academy of Sciences to advise the nation on medical and health issues. Members are elected by their peers for distinguished contributions to medicine and health. Dr. Victor J. Dzau is president.

The three Academies work together as the **National Academies of Sciences, Engineering, and Medicine** to provide independent, objective analysis and advice to the nation and conduct other activities to solve complex problems and inform public policy decisions. The National Academies also encourage education and research, recognize outstanding contributions to knowledge, and increase public understanding in matters of science, engineering, and medicine.

Learn more about the National Academies of Sciences, Engineering, and Medicine at **www.nationalacademies.org**.

The National Academies of

SCIENCES · ENGINEERING · MEDICINE

Consensus Study Reports published by the National Academies of Sciences, Engineering, and Medicine document the evidence-based consensus on the study's statement of task by an authoring committee of experts. Reports typically include findings, conclusions, and recommendations based on information gathered by the committee and the committee's deliberations. Each report has been subjected to a rigorous and independent peer-review process and it represents the position of the National Academies on the statement of task.

Proceedings published by the National Academies of Sciences, Engineering, and Medicine chronicle the presentations and discussions at a workshop, symposium, or other event convened by the National Academies. The statements and opinions contained in proceedings are those of the participants and are not endorsed by other participants, the planning committee, or the National Academies.

For information about other products and activities of the National Academies, please visit www.nationalacademies.org/about/whatwedo.

STEERING COMMITTEE

Leah Jamieson (NAE), *Chair*, The John A. Edwardson Dean, College of Engineering, and Ransburg Distinguished Professor of Electrical and Computer Engineering, Purdue University
Stephanie Adams, Dean, Frank Batten College of Engineering and Technology, Old Dominion University
Marilyn Barger, Principal Investigator and Executive Director, Florida Advanced Technological Education Center of Excellence (FLATE), Hillsborough Community College
Steven Brown, Professor, Counseling Psychology Program School of Education, Loyola University
Don Giddens (NAE), Dean Emeritus, College of Engineering, Georgia Tech
Asad Madni (NAE), Former President, Chief Operating Officer, and Chief Technology Officer, BEI Technologies Inc.
Tom Perry, Director, Engineering Education, ASME
Anne Spence, Professor of the Practice, University of Maryland, Baltimore County
John Wall (NAE), Vice President and Chief Technical Officer (ret.), Cummins, Inc.
Gregory Washington, The Stacey Nicholas Dean of Engineering, University of California, Irvine

Wendy Cowan, AAES Observer

NAE Project Staff

Proctor Reid, Director of Programs
Kenan Jarboe, Senior Program Officer
Carl Anderson, Program Coordinator
Michael Holzer, Senior Program Assistant
Jason Williams, Senior Financial Assistant (until May 12, 2017)

Acknowledgments

This Proceedings of a Workshop was reviewed in draft form by individuals chosen for their diverse perspectives and technical expertise. The purpose of this independent review is to provide candid and critical comments that will assist the National Academy of Engineering in making each published proceedings as sound as possible and to ensure that it meets the institutional standards for quality, objectivity, evidence, and responsiveness to the charge. The review comments and draft manuscript remain confidential to protect the integrity of the process.

We thank the following individuals for their review of this proceedings:

Monitor:
John L. Anderson, Distinguished Professor of Chemical Engineering, Illinois Institute of Technology

Reviewers:
Stephanie G. Adams, Dean, Batten College of Engineering and Technology, Old Dominion University
Nicholas L. Desport, Colonel (Ret.), USAF, Deputy Executive Director and Director of Programs, Membership and Post Operations, Society of American Military Engineers
Michael Gross, Associate Professor, Department of Chemistry, Wake Forest University

Elizabeth Holloway, Assistant Dean of Engineering for Undergraduate Education and Director of Women in Engineering Program, College of Engineering, Purdue University
Harriet B. Nembhard, School Head and Eric R. Smith Professor of Engineering, Department of Mechanical, Industrial and Manufacturing Engineering, Oregon State University
James M. Tien (NAE), Distinguished Professor and Dean Emeritus, College of Engineering, University of Miami
Phillip Westmoreland, Professor, Department of Chemical and Biomolecular Engineering, North Carolina State University, and Director, Southeast Regional Manufacturing Center, Clean-Energy Smart Manufacturing Institute™

Although the reviewers listed above provided many constructive comments and suggestions, they were not asked to endorse the content of the proceedings nor did they see the final draft before its release. The review of this proceedings was overseen by **Janet Hunziker**, senior program officer with the National Academy of Engineering. She was responsible for making certain that an independent examination of this proceedings was carried out in accordance with standards of the National Academies and that all review comments were carefully considered. Responsibility for the final content rests entirely with the rapporteur(s) and National Academy of Engineering.

In addition to the workshop presenters, moderators, and session rapporteurs, we especially thank those individuals who presented so succinctly during the lightning round: Karin Anderson, Nichol Campana, Chris Ciuca, Bob Fine, Jim Hill, Libby Jones, Glenda La Rue, Larry Larson, Aisha Lawrey, Cathy Leslie, Bill Mahoney, Leslie Nolan, Melissa Prelewicz, Karl Reid, Randi Rosebluth, Constance Thompson, Kris Ward, Bill Wepfer, and Phil Westmoreland.

Contents

Appendixes

1

Introduction, Background, and Organization of the Report

Engineering professional societies in the United States are engaged in a wide range of activities involving undergraduate education. However, these activities generally are not coordinated and have not been assessed in such a way that information about their procedures and outcomes can be shared. Nor have they been assessed to determine whether they are optimally configured to mesh with corresponding initiatives undertaken by industry and academia. Engineering societies work largely independently on undergraduate education, leaving open the question of how much more effective their efforts could be if they worked more collaboratively—with each other as well as with academia and industry.

The National Science Foundation (NSF) has long been interested in the role of professional societies in all levels of science, technology, engineering, and mathematics (STEM) education. To explore the potential for enhancing societies' role at the undergraduate level, it asked the National Academy of Engineering (NAE) to hold a workshop on the engagement of engineering societies in undergraduate engineering education. The workshop was held January 26–27, 2017, in Washington, DC. (The workshop agenda is in appendix A.)

In her introduction at the workshop, Leah Jamieson, chair of the workshop steering committee and the John A. Edwardson Dean of Engineering and Ransburg Distinguished Professor of Electrical and Computer Engineering at Purdue University, noted that the goal was to increase "mutual understanding of what engineering societies are doing and can do."

Societies are involved in education in many ways, Jamieson said. A long-time active member of the Institute of Electrical and Electronics Engineers, she has worked throughout her career on involving professional societies more effectively in undergraduate engineering education. She noted that one obvious point of connection between professional societies and engineering education is through accreditation, including the criteria established by the Accreditation Board for Engineering and Technology (ABET). She added, "ABET is a well-known mechanism of engagement, but our goal is to look beyond ABET."

Jamieson identified three questions to be addressed by workshop participants:

1. How to promote more effective roles for societies in the education of undergraduates.
2. How to increase mutual understanding between societies and academic institutions.
3. How to foster collaboration.

Creating change requires thinking creatively about what engineering societies can do separately and together, what they can learn from each other, how they can serve both their own objectives and those of students, and how much more they can do collaboratively than on their own, Jamieson said. Larger and more extensive partnerships can tackle bigger problems, she pointed out, including systemic challenges.

Proctor Reid, director of the NAE Program Office, observed that the NAE is especially well placed to convene a broad range of organizations and individuals with a stake in undergraduate engineering education. The members of the Academy are leaders in industry (40–50 percent have significant industry experience), academia, government, nonprofit organizations, and other sectors. NAE committees and other activities involve both members and other volunteers with wide-ranging backgrounds and viewpoints. In this way, he said, the Academy seeks "to marshal the expertise of the nation's eminent engineers to provide independent advice on matters that involve engineering and technology."

BACKGROUND TO THE WORKSHOP

To establish a foundation for the workshop discussions, a survey, interviews, website review, and literature review were conducted beforehand and summarized in the opening session.

Survey Results and Interviews

A survey of 121 engineering-related societies resulted in 58 responses, after which 30 interviews were conducted with representatives of the responding societies.[1] The results reveal that the respondents are engaged in a range of education activities that target a variety of audiences (see appendix B). Goals of these activities include leadership development, diversity enhancement, and student retention. "Somebody is working toward almost any goal you could imagine," reported Jamieson. Activities include student chapters, faculty development, continuing education, certification, and member communication.

Undergraduate education is not the top priority of most of the societies surveyed but ranks highly across many: 50 percent of the small and medium-sized societies rank undergraduate students as a high-priority audience for their education efforts, as do 65 percent of the large societies and 83 percent of the extra-large societies.[2] Most of the societies focus on creating greater awareness of the importance of engineering in general or of a particular engineering discipline.

Asked about barriers to change, survey respondents identified communications, resistance to change, and limited time, funding, and other resources. A subanalysis revealed that affinity societies are more likely to report barriers than are disciplinary societies.[3]

---

[1] The 121 societies represented an attempt to reach out to all engineering-related societies in the United States (some responded that engineering was not their only priority). Representatives of the societies included executive directors, presidents, chief executive officers, chairs of boards of directors, staff officers, and directors of education or university programs.

[2] Small societies are those with a membership of under 1,000; medium with a membership of 1,000 to 9,999; large 10,000 to 49,999; and extra-large with over 50,000 members.

[3] Disciplinary societies are those organized around a specific professional discipline, such as the American Society of Civil Engineers (ASCE) and Society of Manufacturing Engineers (SME). Affinity societies are those organized around a specific demographic group or nondisciplinary interests, such as the Society of Women Engineers (SWE), National Society of Black Engineers (NSBE), and Engineers Without Borders (EWB).

The responding societies do not devote many resources to evaluation and assessment of their educational activities, beyond simply counting the number of participants in an activity. But they expressed interest in evaluating the impact of their programs in more coherent, rigorous, and longitudinal ways.

A majority of societies (85 percent) consider themselves leaders in the educational field. Yet half rate themselves as having low or "some" capacity to plan and implement education work; 38 percent report that their capacity is high, and 12 percent that it is very high.

Societies' efforts to disseminate their practices are largely limited to their own memberships (via conferences, newsletters, meetings, and other means), but respondents indicated that they are interested in sharing practices beyond their membership.

Finally, societies report a number of gaps in engineering education that educators and engineering organizations could address, including precollege education, faculty preparation for teaching, design education, and two-year preparation programs.

Jenifer Helms of Inverness Research Associates, which conducted the survey, pointed out that close to 90 percent of respondents said that they are partnering or collaborating with other societies in some way, and more than three quarters partner with other organizations. "There's certainly a lot of capacity and desire to engage in partnerships and opportunities to learn how these partnerships and collaborations are established and sustained," she said.

Literature and Website Review

The literature review revealed that formal documentation of the influence of engineering societies on undergraduate engineering education is limited. The American Society of Civil Engineers' (ASCE) Body of Knowledge project[4] is one outstanding exception, noted Jamieson, as is the work of ABET. The literature also provides some information on student chapters, the roles of societies in promoting diversity, and informal education activities such as community service projects.

A review of 122 websites associated with engineering societies revealed that almost all have some form of direct engagement with students—if not necessarily with engineering education—mainly through student mem-

[4] See www.asce.org/civil_engineering_body_of_knowledge.

berships and student chapters.[5] In addition, about one quarter offer some form of financial assistance such as travel grants to meetings or scholarship support. More than three quarters have some form of indirect engagement through technical publications, standard setting, and educational research articles and briefs. However, the websites of societies contain little or no evidence of how these activities relate to the effectiveness of the societies' engagement with students, such as student retention or preparedness to enter the workforce.

ORGANIZATION OF THE WORKSHOP PROCEEDINGS

The material presented at the workshop is organized into six chapters. After this introductory chapter, chapter 2 describes the "ecosystem" for engineering education presented by Elliot Douglas, program director in the NSF Division of Engineering Education and Centers. His talk established a context for the workshop discussions.

Chapter 3 summarizes the keynote address on the state of engineering education by Darryll Pines, dean and Nariman Farvardin Professor of Aerospace Engineering at the University of Maryland's Clark School of Engineering.

Chapter 4 provides an account of the presentation by Barbara Bogue and Betty Shanahan, principal investigator and co–principal investigator of the Society of Women Engineers' (SWE) Assessing Women and Men in Engineering project, about the importance of effective assessment.

Chapter 5 presents brief summaries of educational activities being conducted by engineering societies. These "lightning" presentations are divided in two categories: examples of what societies are doing and of how they are doing what they do (i.e., overcoming concerns, problems, and barriers).

Finally, chapter 6, "From Analysis to Action," compiles the suggestions of subgroups that met during the two days of the workshop to discuss specific issues and propose ways to enhance the engagement of engineering societies in undergraduate engineering education.

[5] The website review covered the 121 societies identified for the survey plus ABET.

2

An Ecosystem Perspective

Future engineers take many pathways through the educational system, said Elliot Douglas, program director in the Division of Engineering Education and Centers at the National Science Foundation. They are subject to many influences and bring their own characteristics and educational trajectories to their undergraduate experiences. For these reasons, thinking of engineering education as a leaky pipeline can be misleading. A better metaphor, he said, is that of a much larger ecosystem—of which engineering education is a part—characterized by myriad proximal and distal interactions among a large number of actors and influences.

THE PROFESSIONAL FORMATION OF ENGINEERS

Professional societies are a prominent part of this ecosystem. Their influence on students is often indirect, observed Douglas, although in some cases they work directly with students. But they influence many other parts of the system, and these other parts can influence undergraduate engineering education.

Because the influence of engineering societies is often indirect, NSF has not previously focused directly on their role in undergraduate engineering education, Douglas noted. However, the development of a new NSF initiative called the Professional Formation of Engineers has expanded the foundation's interests from engineering education narrowly defined to the formation of an engineering identity, which has in turn increased attention on professional societies.

The professional formation of engineers encompasses the formal and informal processes and value systems by which people become engineers, Douglas explained. Elements include:

- introduction to the profession at any age
- acquisition of deep technical and professional skills, knowledge, and abilities in both formal and informal settings and domains
- development of outlooks, perspectives, and ways of thinking, knowing, and doing
- development of identity as an engineer and its intersection with other identities
- acculturation to the profession, its standards, and its norms.

THE ROLES OF PROFESSIONAL SOCIETIES

Professional societies are involved in all these elements, Douglas observed. As one example, he pointed to recent discussion of the "T-shaped" engineer who combines both breadth of knowledge and depth of expertise (or the ability to apply knowledge across situations as well as functional/disciplinary skills).

To develop breadth, students need an understanding of how their field interfaces with other fields. They also need skills such as communication, critical thinking, metacognition, and leadership, Douglas said. Students with these abilities have the potential to become adaptive experts, able to restructure knowledge depending on the situation. Professional societies can help establish the norms and expectations that build such expertise.

Societies also influence thinking about diversity, including not just the representation and presence of diverse people but the inclusion of diverse perspectives, knowledge of different social identity groups, and considerations of social justice (including power, privilege, and oppression). Douglas characterized engineering as a sociotechnical profession, not just a technical profession, involving noncognitive factors, such as motivation and self-regulation, as well as cognitive factors. Educational success often depends on social connections with communities, families, and social groups. (Lack of social capital is also why first-generation college students can face higher barriers to success, Douglas noted.) These are all factors in the broad ecosystem of engineering education.

The traditional way of doing engineering has been to solve specific problems. This approach to engineering may be why longitudinal studies have found that belief in the importance of engineering's impact on society gradually diminishes among engineering students.

Some educational programs are countering this approach and trend by focusing on the humanitarian and social justice aspects of engineering. Furthermore, many opportunities for broader access exist, for example through online education and the inclusion of engineering in the Next Generation Science Standards.

"Again, professional societies are part of [all these opportunities] because you have the ability to impact the field broadly, not just within a single institution or a single classroom," said Douglas.

A NEW FUNDING MODEL

Scalability cannot be ignored, Douglas said. The traditional funding model has been to support principal investigators in developing innovations for their classrooms. And the traditional dissemination strategy has been to publish in peer-reviewed journals, create websites, and give workshops at meetings such as those of the American Society for Engineering Education.

The new funding model supports large integrated efforts such as NSF's REvolutionizing engineering and computer science Departments (RED) program and the Improving Undergraduate STEM Education Framework. A new dissemination strategy is to consider models of change and the creation of national cohorts of exemplars. "You don't start from 'I want to do this activity,'" said Douglas. "You start from 'I want to make this cultural change.' That's a very different way of thinking."

A goal of the workshop was to inculcate this different way of thinking, Douglas concluded. "Let's think about how to not just cross-fertilize but cross-collaborate and create these larger partnerships that can work more broadly and at a larger scale to impact the engineering education field. What we want is broad, radical change in engineering education."

3

The State of Engineering Education

Every organization and every country on the planet is hungry for talent, said NAE president C. D. Mote, Jr., in introducing keynote speaker Darryll Pines, dean and Nariman Farvardin Professor of Aerospace Engineering at the University of Maryland's Clark School of Engineering. At the same time, engineers and other workers are changing jobs more often as the pace of change in society accelerates. Both trends cast a spotlight on engineering education and on the ways professional societies are involved with this education.

THE ATTRACTION OF ENGINEERING

Engineering is one of today's hottest professions, Pines began. Since the year 2000, US enrollment in engineering has gone from less than 400,000 to more than 700,000. This surge in enrollment has transformed undergraduate engineering programs. "It's an exciting time to be an engineer," Pines said.

Increasing Enrollment

The increase in enrollment has encompassed both men and women and all ethnic groups, although women and many ethnic groups remain underrepresented in engineering compared with their representation in the general population. For example, about 23 percent of US engineering undergraduates (not including foreign students) are currently women, an increase from historical levels of below 20 percent. "We are becoming more balanced," said Pines. "We are still not there by any stretch of the imagination, but the demographics are slowly moving."

The number of engineering bachelor's degrees awarded increased from fewer than 80,000 in 2005 to more than 115,000 in 2015. This has put a greater burden on faculty members, Pines pointed out: The average number of engineering students per tenure track faculty member rose from about 17 to about 25.

Top Salaries

Notwithstanding the increase in degree production, engineers are still getting good-paying jobs, Pines observed. Of the top 15 majors by salary, according to the website www.payscale.com, 11 are in engineering fields (the other four are in actuarial science, computer science and mathematics, physics and mathematics, and applied mathematics). Early career pay for these engineering fields ranges from $63,000 to almost $100,000, and midcareer pay is between $108,000 and $172,000.

What Is Engineering?

Despite success in the field, engineers traditionally have had a hard time defining what they do, Pines pointed out. The Accreditation Board for Engineering and Technology (ABET) defines engineering as "the profession in which a knowledge of the mathematical and natural sciences gained by study, experience, and practice is applied with judgment to develop ways to utilize, economically, the materials and forces of nature for the benefit of mankind." Pines said that he prefers the more succinct definition offered by NAE president Dan Mote: "Engineers create solutions serving the welfare of humanity and the needs of society." The four words *creation*, *solutions*, *humanity*, and *society* together create a value proposition for engineering, Pines said, and offer a way to communicate what engineering is to the public and to students.

Historical Perspective

Engineering and engineering education have changed radically through history and are continuing to change today, Pines reported. The first three schools in the United States to offer engineering education were the US Military Academy, which modeled its engineering curriculum after the École Polytechnique in France; an institution now known as Norwich University in Vermont, which began with instruction in civil engineering; and Rensselaer Polytechnic Institute in New York. Engineering education greatly expanded as part of the Morrill Acts of 1862 and 1890, which created the land-grant colleges and universities and the historically black colleges and universities.

Throughout the 19th century, engineering education was largely focused on practice, including shop and foundry skills, technical training, and manufacturing. This is the period that saw the creation of some of the first professional societies—the American Society of Civil Engineers (1852), American Society of Mechanical Engineers (1880), and American Society for Engineering Education (1893).

In the first half of the 20th century, the emphasis shifted from practice to theory and science, driven in part by World Wars I and II. ABET was established in 1932 to help set standards for the engineering curriculum, along with such societies as the American Institute of Chemical Engineers (1908) and the precursors to the Institute of Electrical and Electronics Engineers and the American Institute of Aeronautics and Astronautics.

In the second half of the 20th century, an emphasis on engineering design swung the pendulum back from theory toward practice and hands-on engagement, said Pines, with a focus on project-based learning, hands-on and applied work, ethical reasoning, professional development, and industry collaboration.

Modern Engineering Education

Most recently, engineering education has emphasized research, complex systems, pedagogy, active learning, service learning, teamwork, online education, virtual laboratories, communication, creativity, leadership, global contextual analysis, innovation, and entrepreneurship. Furthermore, new departments of engineering education mark "a paradigm change," noted Pines, where research on learning is being used to improve engineering education, and faculty are being rewarded not only for their technical research but for their contributions to teaching and learning.

Many forces have been driving change in engineering education, said Pines, including engineering college and departmental or program advisory boards, professional societies, the National Academy of Engineering, the National Science Foundation, industry and private foundations, ABET, and advances in research, facilities, and technology. For example, the NAE reports *The Engineer of 2020*[1] and *Educating the Engineer of 2020*[2] are among a series of reports demanding change in engineering to serve the welfare of humanity and society.

IMPLICATIONS FOR ENGINEERING EDUCATION

Modern engineering is increasingly complex, Pines pointed out, and increasingly tied to US economic competitiveness and issues of great societal importance. Data science and analytics are accelerating research and steering it in new directions. The development of engineering systems typically reflects and draws on the convergence of the natural sciences, the social sciences, medicine, management, the humanities, and other fields. Engineers need to be "more systems oriented," said Pines, "to develop large systems and have people be able to model them."

Curricula in engineering education have evolved to reflect these changes, said Pines. The overall goal has been to create a multiyear, vertically integrated, hands-on, active learning experience.

The first-year experience may involve design, team building, novel classroom environments (such as maker spaces or flipped classrooms), and work involving innovation and entrepreneurship. Second- and third-year engineering courses (which is when community college transfer students enter) may involve leadership and business management, international experiences, and internships. Senior capstone experiences can include mentoring younger students, making links to industry and graduate education, and becoming involved with professional societies.

The engineering school at the University of Maryland, for example, with seven four-year and two-year collaborators, has created the Keystone

---

[1] NAE [National Academy of Engineering]. 2004. *The Engineer of 2020: Visions of Engineering in the New Century.* Washington, DC: The National Academies Press.

[2] NAE. 2005. *Educating the Engineer of 2020: Adapting Engineering Education to the New Century.* Washington, DC: The National Academies Press.

program to respond to calls for active hands-on learning. It is designed to transform the first- and second-year experience by providing incentives to instructional faculty, improving facilities, organizing team competitions, encouraging the use of undergraduate teaching fellows, offering peer mentoring, and enhancing the involvement of professional engineers as mentors, advisors, and reviewers. This program has helped boost the six-year completion rate at the school to 75 percent, compared with a national average of 59 percent.

Throughout their undergraduate years, students at Maryland have opportunities for individual learning experiences such as "hackathons," student competitions or challenge prizes, service learning, and community engagement.

- A University of Maryland group was one of the teams selected to participate in a hyperloop design competition organized by Tesla founder Elon Musk.
- A program called StartupShell has resulted in startup companies that are selling consumer 3D printers and recovering leftover food from university dining halls to feed the hungry.
- A solar decathlon brought together people from business, science, public policy, architecture, and other fields to design an energy-efficient house that was sold to a leading energy firm to use as a model for new technology.
- A course on engineering for social change, in which students work with a local community to design an innovative solution to a problem and pitch their ideas for foundation funding, resulted in an educational program that integrates gardening, cooking, and nutrition in the curriculum of local schools.

Innovations in engineering education can encompass the K–12 level as well. A University of Maryland hackathon for middle and high school girls was designed to get more girls interested in computer science and engineering.

Another approach to engineering education that reflects the needs of the 21st century is based on the 14 Grand Challenges for Engineering (www.engineeringchallenges.org) identified by an NAE committee in 2008. The Grand Challenges Scholars Program (www.grandchallengescholars.org) guides students, through curricular and extracurricular activities, to gain skills in five required areas: research related to a Grand Challenge, a multidisciplinary experience, exposure to the global dimension of a Grand Challenge, entrepreneurship, and service learning. More than 40 universities in the United States and abroad have adopted the program, and another 80 have committed to participating.

These disparate approaches embody common desires for engineering education and engineering students, said Pines:

- the inculcation of engineers as problem definers as well as problem solvers
- the development of engineers who are better able to straddle uncertainty, risk, disciplines, cultures, ethics, and evolving technologies
- engineers who are prepared for creativity, innovation, business management, entrepreneurship, and public policy leadership
- engineers who have stronger application skills without losing theoretical strength.

TODAY'S ENGINEERING STUDENTS

The millennial and Gen Z students in college today are different from past generations of students, Pines said. They have different work ethics, career expectations, management styles, and knowledge of technology. "They want to see significant change in their lives, and they want it quickly," he said. "They want to work on projects that inspire and have social impact." They are already making a difference in the workforce, and that influence will grow.

One important aspect of young workers is that they are digital natives. They are always online, socially conscious, and socially connected. They understand blogs, social networks, mobile devices, and online tools. "They are confident, they are connected, and they are open to change," said Pines. "They want to make a difference with their knowledge and with the skills they get from our schools."

This comfort with technology is helping to drive a new approach to education, one that includes blended learning with online lectures, automated assessments of student performance, flipped classrooms with peer-to-peer and instructional coaches, and massive online open courses (MOOCs) to enhance learning outcomes.

"Learning is in a transition," said Pines. It is increasingly self-paced, self-serviced, virtual, and on demand. Technologies such as content capture, online laboratories, learner profiles, and e-portfolios are the future of instruction and learning. Virtual laboratories make it possible to have students go to an online location, run an experiment, get the data back, and report on those data even if they were not physically present in a lab.

Higher education got an early taste of radical changes in the ecology and economics of education with the advent of MOOCs in the late 1990s and early 2000s. Today, a new generation of online courses, such as those made available through Udacity, Coursera, edX, and the Khan Academy, are providing new capabilities in an era of economic pressures and a social readiness to embrace distributed relationships between students and instructors. For example, Arizona State University and the State University of New York (SUNY)–Oswego are offering the first two ABET-accredited online undergraduate engineering programs in electrical and computer engineering.

But residential universities, especially in STEM fields, are not going away, Pines continued. Although many engineering courses could be taught online, design is a creative activity that probably needs to be taught in an integrated environment. Hands-on laboratory experiences remain crucial, even if some laboratory experiences can occur online. An optimal education still requires interactions between students and teachers, and no professor can do that for thousands of students in an online course.

Pines also pointed out, in response to a question, that much more is needed to increase the representation of women in engineering. Engineering for social change—for example, through the Engineers Without Borders program—is particularly engaging to women.

Engineering programs also need to work with K–12 engineering education so that fewer girls lose interest in STEM subjects in high school, middle school, and even elementary school. "Engineering has to become part of the core education for K through 12, not a fringe topic," Pines said. For example, the Engineering Is Elementary program developed by the Museum of Science in Boston focuses on teaching engineering habits of mind to elementary students, getting them thinking about design and connecting them to creativity.

THE ROLE OF PROFESSIONAL SOCIETIES

Professional societies have critical roles to play as this new engineering education paradigm emerges, said Pines.

- By fostering industry-university collaborations, they can help define real-world challenges that require innovative contributions from universities.
- They can help organize student-directed, hands-on learning such as annual competitions, and they can provide advice, guidance, and critical review for capstone educational experiences.
- They can further competency-based education, with companies helping to define particular competencies for students to acquire in university courses. One possibility, for example, would be a course series designed by professional societies to teach skills and content that students need, such as standards or ethics.

A new normal is evolving in engineering education, Pines said in summary. The value proposition that centers on creation, solutions, humanity, and society is creating a greater emphasis on hands-on and experiential learning opportunities in the context of current and future societal challenges. Professional societies can play a critical role in this new paradigm by:

- connecting engineering education to real-world practice and solutions
- serving as design team reviewers, mentors, advisors, and educators
- creating challenge projects to advance technology and skills
- providing opportunities for international and service learning
- serving as ambassadors to the profession through outreach to K–12 education.

4

The Need for Effective Assessment

Engineers strive to apply good practices in their profession, said Barbara Bogue and Betty Shanahan, principal investigator and co-principal investigator of the Society of Women Engineers' Assessing Women and Men in Engineering project, yet they often fail to do so when engaged in outreach. The survey and interviews conducted before the workshop revealed that assessment is a critical but often missing influence on professional societies' outreach efforts. Bogue and Shanahan made the case that effective assessment should be the basis for all engineering outreach initiatives.

AN ASSESSMENT-BASED FRAMEWORK

An assessment-based framework aligns collaboration and outreach practices with typical engineering design and project management practices. It involves identifying audiences, specifying goals and objectives, and defining the metrics and data to be used in measuring outcomes. An assessment-based framework is a core tool for successful initiatives, Bogue and Shanahan said.

An assessment-based framework can apply at a meta level and at the level of specific actions, Shanahan explained. For both an overall collaboration and individual societies, it can ensure that goals are relevant to the mission and goals of each partner as well as the collaboration. It can have built-in measures of the impacts of activities and create a continuous improvement platform that serves the overall outreach programming.

Bogue and Shanahan illustrated some assessment issues by recounting the outcomes of an outreach program Bogue offered at Pennsylvania State University. A one-week residential engineering camp had the objective of recruiting high school girls into the engineering profession. The program specifically recruited girls who did not plan to become engineers. The program emphasized hands-on projects led by role models, and 42 girls participated. Postevent survey results indicated that all 42 participants were enthusiastic about the event. Before the program, 40 of the participants said they did not plan to study engineering; after the program the same number said they wanted to become engineers. In addition, all 12 of the participants who were high school seniors said they planned to apply to engineering at Penn State.

Postcamp tracking revealed, however, that only two participants followed through in applying to Penn State, and only one was accepted. Furthermore, the camp was expensive to administer—about $1,400 per girl—and a time analysis revealed that only about a quarter of their time at the camp was spent on engineering activities. "In the postreview, [the program] failed on a lot of different points," said Bogue.

Based on these findings, the camp was radically redone. It moved from an overnight camp to a day camp with modules that had a different interdisciplinary focus each day. Young women could come to the camp on one day, two days, three days, or five days, and the camp served many more girls—more than 300 as opposed to a maximum of 60 in the overnight camp.

Moreover, the revised camp design had a much greater concentration on objectives and outcomes, and better pre- and postcamp assessments produced more relevant data. Ninety percent of time was now spent on engineering-related activities. Resources also were used more efficiently, with a cost of $142 per girl per day. In addition, the organizers avoided areas of nonexpertise: "no more slumber parties," said Bogue.

Bogue identified several lessons from this experience:

- Participants having fun is a success indicator only if the only goal is fun.
- Poor or incomplete data can lead to wrong overall evaluations and decision making.

- Surveys that do not ask the right questions produce the wrong answers.
- Adding data through postassessments can lead to more accurate evaluations.

CHALLENGES TO EFFECTIVE OUTREACH

Shanahan noted that engineers would never approach a problem the way we often approach outreach. Companies would not begin product development without reviewing relevant technologies, determining customer needs, and establishing product goals and objectives around costs, performance, and safety. So why does outreach by engineering societies so often fail to incorporate standard engineering design? One reason is a failure to identify and serve an intended audience. In practice, the de facto audience often becomes the member volunteers or funders rather than the kids that the activity is supposed to reach. As Bogue put it, "Are we going to make Joe unhappy because we're not offering his camp the tenth year in a row?"

Another reason for unsuccessful outreach is a failure to define the value added for every partner, including the volunteers. Limits on human and financial resources are a barrier, as are actions that belie the goal of an outreach program. For example, if a goal is to attract underrepresented students to engineering, why are members of a society so often unaware of outreach programs and why are the programs poorly resourced?

APPLYING WHAT ENGINEERS PRACTICE

An assessment-based framework can help meet these challenges, said Shanahan and Bogue. The framework is based on an agreed-upon, shared, and overarching goal. With the goal established, measurable objectives should be determined that will fulfill that goal, they said. These objectives describe what the initiative will achieve rather than describing an activity, and they create the foundation for planning, assessment, and continuous improvement.

The next step is to leverage resources and define initiatives, using research to inform choices and designing initiatives based on the goal and objectives, not the other way around. This research can come from many different sources, including the social sciences, and can be informed by the practices of other organizations. Shanahan mentioned TED talks as a particularly useful source; they can be distributed to volunteers as 15-minute videos that capture the essence of research results.

A data collection plan with defined metrics needs to be created, Shanahan and Bogue continued. If the goal is to reach underrepresented populations, count the number of people who are reached. Use before and after questions to assess changes in knowledge, interest, skills, or confidence. Surveys, formal observations, and formal interviews are all ways of gathering data; anecdotal information may be interesting or useful but it is not trackable, comparable, or objective. Longitudinal data are the gold standard, but they are often difficult to gather.

Time needs to be scheduled for data collection and analysis, with online tools to facilitate these steps. Collaboration in data collection and analysis can ease obstacles to sharing information and provide information that is useful for shaping other initiatives.

Assessment can be affordable if it is integrated into an overall plan and scheduled from the beginning. Bogue and Shanahan offered several pointers:

- Identify a volunteer who enjoys working with data, enlist the educational arm of a society in data collection and analysis, centralize data, use off-the-shelf resources, and create a bank of common surveys and tools.
- Recruiting, selecting, and training volunteers is crucial to success, and they need to understand the goals and objectives, the assessment-based approach, and outcome metrics.
- Break down tasks by interests and skills so that no role is too big and a small number of volunteers do not carry an intensive load.
- Enlisting experts at developing and analyzing outreach initiatives can be much more effective than training nonexperts to serve in a key role.

Once data are collected and analyzed, they can be used for continuous improvement. "It's not just what we know," said Shanahan. "It's what we do with what we know. Use the results to say, 'This went well, this didn't go well. How do we change? How do we improve? How do we enhance?' and invest in making those changes." Initiatives that do not work are more expensive than good assessment.

Finally, tell the story to recruit participants, motivate volunteers, engage and convince board members and sponsors, enhance and expand collaborations, and share positive and negative lessons. A good story, backed by solid data, can enable programs to be scaled up to have greatly magnified impact.

THE NEED FOR LEADERSHIP

Research in change management shows that change does not happen without leadership driving it, said Shanahan. "You are the ones who are going to make change happen. It's not going to be your volunteers. . . . As someone who was . . . a society leader, I know it's not easy to say to your volunteers, 'You're going to have to do more work' or 'We've got some bad news here, and we need to respond to it.' It is challenging. But our mission as engineering societies [doing] outreach programs is not to create fun for volunteers. It's to have effective outreach."

5

Engineering Society Activities

In two sets of "lightning rounds" on the first day of the workshop, representatives of engineering societies presented brief descriptions of the activities of their organizations related to undergraduate engineering education. The first set of presentations focused largely on *what* societies do; the second focused largely on *how* societies do what they do, though some overlap with the previous presentations was inevitable.

WHAT SOCIETIES DO

Developing Partnerships for Innovation in Education

SME Certified Manufacturing Technologists *(Kris Ward)*

The Society of Manufacturing Engineers (SME) Certified Manufacturing Technologists (CMfgT) certification benefits entry-level manufacturing technologists and experienced manufacturing engineers without other credentials. The end-of-program assessment is used in both two- and four-year programs in engineering technology, manufacturing technology, manufacturing engineering, and mechanical engineering.

The CMfgT is built on an industry-driven competency framework and on a body of knowledge that reflects the entry-level requirements for engineering and manufacturing technologists. A number of optional resources are provided through the academic market to help students prepare for the assessment. The exam is not only an effective student evaluation but also a tool that provides insight into academic program performance. Individual and group reports allow instructors, deans, and department heads to assess student knowledge and identify gaps in curriculum against a standard body of knowledge. These measures help to continually improve the program so that graduating students meet industry-driven requirements.

The assessment supports ABET accreditation requirements, and students who pass the exam also get an industry credential that differentiates them to employers. The program's continuing education requirement gives schools the opportunity to engage with students after graduation. The CMfgT offers an industry-driven alternative to the Fundamentals of Engineering (FE) exam.

For more information visit www.sme.org/cmfgt.

SWE Collegiate Leadership Institute *(Randi Rosebluth)*

The objective of the Collegiate Leadership Institute (CLI) is to provide cutting-edge leadership and career development training for college and graduate students in technology and engineering. The school-year-long program accommodates 125 students and serves as an ongoing dynamic research platform for Society of Women Engineers (SWE) college leaders. In 2018 the CLI will expand to local conferences and a conference in India.

The CLI is modeled around pre- and postconference online learning. The institute includes a three-day live event, online engagement, webinars, mentors, individual coaching, a $150 stipend for transportation, and networking. Small groups are also offered career and professional coaching. The purpose of the program is to provide tools for the workplace, develop a leadership pipeline that enables women to take on roles in SWE within 5–15 years, and foster lifelong involvement with SWE.

For more information visit http://societyofwomenengineers.swe.org.

ASME Faculty/Industry Standards Experts Teams *(Bill Wepfer)*

About 10 years ago the American Society of Mechanical Engineers established a task force in the spirit of the NAE *Engineer of 2020* Initiative to determine how the mechanical engineering profession can make its curriculum relevant for students over the next 20 years. Department heads, hiring managers, and industry leaders were surveyed to define the most pressing issues. Communication and understanding of codes and standards in undergraduate programs emerged as the strongest theme.

Three task forces were developed to embed standards in the undergraduate curriculum for design track/senior capstone design, mechanics of materials, and fluid mechanics. Each was led by both faculty experts and standards experts. Two primary questions that the task forces hope to address are how to embed codes and standards in a more positive context that facilitates problem solving and economic development, and how to recruit more talented and qualified faculty to help successfully run the program.

For more information visit https://www.asme.org/about-asme/standards/engineering-student-resources.

AIChE Manufacturing Institutes (Phil Westmoreland)

AIChE: The Global Home of Chemical Engineers is participating in a leadership position in two Manufacturing USA institutes, which are an initiative of the US Department of Defense, the National Institute of Standards and Technology, and the Department of Energy, to help foster a cross-disciplinary culture across engineering disciplines. Through its RAPID Manufacturing Institute, AIChE is working with key partners to increase efficiencies and lower capital costs through rapid advancement in process intensification deployment. The Clean Energy Smart Manufacturing Innovation Institute is working to spur advances in smart sensors and digital process controls that can radically improve the efficiency of US advanced manufacturing. Faculty and student involvement is an important part of the institutes' activities.

For more information visit https://www.aiche.org/rapid and https://www.cesmii.org.

Promoting Diversity

NSBE Student Retention Toolkit (Karl Reid)

Despite the high demand for engineers, only about one third of African Americans who start on the engineering track complete the program in six years. The National Society of Black Engineers (NSBE) recently announced a 2025 strategy to triple the number of African American engineers produced by colleges and universities. To fulfill that goal, NSBE is working with colleges of engineering to increase the success rate of African American students. The program, with support from ExxonMobil, first recognized exemplar institutions that produce above-average numbers of underrepresented minority engineers. Through literature reviews and interviews about specific programs at these institutions, nine engagement strategies were identified in a white paper: institutional leadership, summer bridge programs, collaborative learning and living environments, early alert systems, facilitated study groups, faculty development, scholarships, self-efficacy, and positive identity development.

In partnership with the American Society for Engineering Education (ASEE), this work led to the publication of the *NSBE Student Retention Toolkit*, available both in print and online, which operationalizes the engagement strategies. This 170-page document is part of NSBE's comprehensive strategy to increase the success of underrepresented minorities in engineering education. The society plans to offer training (as part of preconference workshops) in ASEE and NSBE programs and grants as well as technical assistance to increase the capacity of colleges and universities to implement these strategies.

For more information visit www.nsbe.org.

Transforming Engineering Culture to Advance Inclusion and Diversity
(Glenda La Rue)

Led by the Women in Engineering ProActive Network (WEPAN), the American Society of Mechanical Engineers (ASME), and Purdue University, Transforming Engineering Culture to Advance Inclusion and Diversity (TECAID) is a three-year-long project involving five mechanical engineering departments across the country. The program aims to equip faculty teams of five with the tools needed to create and sustain inclusive department cultures for faculty, staff, and students.

TECAID is an intensive professional development curriculum based on the literature on inclusion, diversity, team building, team-based change leadership, and strategies and processes in academia. The program is run through virtual learning communities, four two-day workshops, subject matter expert consults, and department projects. Goals include academic change, climate and culture improvements, better department leadership policies and practices, and increases in collaboration. TECAID aims to provide faculty with both understanding and comfort with diversity and inclusion concepts, a way to create and implement plans to address these concepts, and confidence and skills in change making.

Departments using the TECAID program have improved in all these areas, particularly through individual growth. Individuals in the program are starting to implement change, and these effects should ripple out to teams, departments, and institutions. TECAID will publish a training model to prepare engineering faculty to lead diversity, equity, and inclusion change.

For more information visit www.wskc.org/tecaid.

Fostering Interdisciplinary Engineering Education

Engineering Competency Model *(Melissa Prelewicz)*

The Engineering Competency Model, a joint initiative of the US Department of Labor and the American Association of Engineering Societies, serves as a guide for the development of professionals and the engineering workforce. Intended for educators, guidance counselors, and students, the model promotes understanding of the skills and competencies needed for a globally competitive workforce. It can inform educators in the development of a competency-based curriculum, assist guidance counselors in the development of resources for career exploration, and help students gain a clear understanding of the skills and abilities needed to not only enter but advance in an engineering profession.

The model uses a five-tier pyramid to depict the key competencies. The first level is personal effectiveness, including interpersonal skills, integrity, and interest in lifelong learning. Tier two, academic excellence, involves communication and writing skills along with science and technology skills. Tier three, workplace skills, focuses on teamwork, creative thinking, and business fundamentals. Tier four is industrywide technical competencies; it concerns not only the fundamentals of engineering but also areas such as professional ethics. Tier five is an opportunity to include discipline-specific competencies, which several societies are developing. A two-minute video, a PowerPoint presentation with a speaker's guide, and a handout are available for faculty, guidance counselors, and others who work with individuals who are entering and working in STEM professions.

For more information visit www.aaes.org/model.

Engineers Without Borders–USA Global Classroom *(Libby Jones)*

The overarching goal of the Engineers Without Borders–USA Global Classroom program is to provide training for members and volunteers that will enable them to work successfully with community partners in America and around the world. The program aims to provide an understanding of the skills and tools needed for development and humanitarian engineering and to prepare the participants to apply these skills and tools in the service of small rural farms in Costa Rica under the supervision of experienced engineers. Costa Rica is the global frontrunner in sustainable development and offers unmatched opportunities to observe sustainable best practices in action. By the end of the course, students produce a successful model of project management, from planning to implementation to monitoring and evaluation.

The curriculum is being improved using assessment feedback from the approximately 50 participants, instructors, and farmers. This input helps guide the development of online modules that can reach a larger population of about 5,000 people.

For more information visit www.ewb-usa.org.

Raising Awareness of Engineering Disciplines

ASM Materials Camps *(Nichol Campana)*

The ASM Materials Education Foundation, the philanthropic division of ASM International, aims to guide young people into materials science and engineering careers to help create a skilled STEM workforce. The foundation's signature program is its Materials Camps for teachers and students. Teachers' camps are week-long, hands-on laboratory experiences that show educators how to use applied engineering techniques in their classrooms. They include an idea-generating workshop introducing teachers to methods that make mathematics and core science principles more appealing and relevant to middle and high school students. Student camps are for high school students with strong abilities in mathematics and science who have completed their sophomore and junior years. Eighty-six percent of student camp participants enroll in science and engineering programs in college.

The more than 70 camps are held at universities and other institutions across the United States as well as in Canada, France, India, and Brazil. Undergraduate students are encouraged to get involved as volunteer mentors, team leaders, laboratory assistants, and occasionally organizers of the camps. The five-day camps can be residential, commuter, or minicamps, and many are held in conjunction with other ASM events.

For more information visit www.asminternational.org/foundation/teachers/teacher-material-camps.

ANS Annual Student Conference (Bob Fine)

American Nuclear Society (ANS) members under the age of 35 are the fastest growing segment of the society's membership. The society has student sections at 52 universities across the country, and about 14 percent of its members are students.

Every year a student section hosts a student conference, representing one of a variety of activities offered to student members. Students plan and organize an educational meeting, conduct sessions, present research findings, hear talks by nonstudents on topics of interest (such as public policy issues), and network, including with recruiters. Participants also include professors, professionals, and job recruiters.

In addition to the annual conference, faculty and guest speakers attend monthly or bimonthly events at these universities.

For more information visit www.ans.org.

SAE Collegiate Design Series (Chris Ciuca)

The SAE Collegiate Design Series aims to facilitate connections between industry and education by providing experiential learning for students. Approximately 120,000 students from pre-K to college age—including 10,000 undergraduate students—participate in SAE STEM programs each year through integrated educational design challenges. The series includes Baja SAE, Formula SAE, SAE Aero Design, SAE Clean Snowmobile, SAE Supermileage, and the new AutoDrive Challenge. Participating students meet and take on the roles of a range of experts in the field, from engineers to business developers and marketing and finance professionals. The series aims to create an environment that prepares students for a career through a university-simulated business-like setting.

SAE's collegiate programming engages teams from all of the top US engineering degree–granting universities (as ranked by ASEE's *Profiles of Engineering and Technology Degree Colleges*), with many participating in multiple SAE Collegiate Design Series competitions.

For more information visit http://students.sae.org/cds/.

HOW SOCIETIES DO WHAT THEY DO: ISSUES, PROBLEMS, AND BARRIERS FACED AND OVERCOME

Establishing Effective Inter-Society Collaborations

The 50K Coalition *(Constance Thompson)*

The American Indian Science and Engineering Society (AISES), Society of Hispanic Professional Engineers (SHPE), Society of Women Engineers (SWE), and National Society of Black Engineers (NSBE) have formed a coalition with one major goal: to graduate 50,000 female, Black, Hispanic, and Native American engineers by the year 2025. Led by the 50K Coalition Leadership Circle, the project has 43 member organizations representing engineering professional societies and schools of engineering that share this focus. The United Engineering Foundation and National Science Foundation have provided half a million dollars in funding.

The coalition uses a collective impact approach, with a common agenda, project plans, and defined metrics. The agenda has six elements: undergraduate support and retention, public awareness and marketing, K–12 support, community college linkages, culture and climate, and funding and financial support. Each organization contributes a project plan that includes an agreed upon matrix and measurable agenda items to serve as its commitment to reaching the goal.

For more information visit http://50kcoalition.org.

Material Advantage Student Program *(Bill Mahoney)*

The Material Advantage Student Program was created for undergraduate and graduate students in materials science, engineering, and other technical engineering programs at universities around the world. It is operated by the American Ceramic Society, Association for Iron and Steel Technology, Materials Information Society, and Minerals, Metals and Materials Society. Each organization shares expenses and revenue, operates programs and procedures on behalf of Material Advantage, and may maintain both Material Advantage–branded programs and their own society programs.

The $30 US membership fee provides full access to all four societies. Students can use their membership to enhance their personal development and to build a career foundation with long-term benefits.

A number of events are executed on behalf of the Material Advantage Program. The culminating event, the Material Science and Technology Conference held every year in the fall, exposes students to a large number of networking, learning, and collaborative opportunities as well as various contests and awards.

For more information visit http://materialadvantage.org.

Bolstering Society-University Collaboration

Tooling U-SME and E-learning *(Kris Ward)*

Tooling U-SME is the workforce development division of SME that connects the local education community, employers, and workforce groups to provide jobs for students and meet employer/workforce needs. The program identifies specific market conditions in order to analyze needs, design a complete program, create a detailed implementation plan, and do follow-through with evaluation and support to ensure the best results and outcomes. It uses flipped classrooms to enable instructors to spend more time on applied and hands-on learning and skill building. It is incorporated in universities, community colleges, and high school bridge programs that award college credit.

Tooling U-SME brings industry-driven e-learning content to higher education that is mapped to the needs of industry competencies, apprenticeship programs, and certifications. Students that use Tooling U-SME classes as part of their education program can port their Tooling U-SME transcript to employers, who can decide the level of credit to provide. This saves the employer significant time and resources when onboarding these new hires, increasing the value provided by the college or university. Using industry-driven resources and creating a connected education-employer community also can help institutions of higher education track placement rates and monitor student postgraduate success.

For more information visit www.toolingu.com.

ASCE ExCEEd Teaching Workshops *(Leslie Nolan)*

The American Society of Civil Engineers' Excellence in Civil Engineering Education (ExCEEd) Teaching Workshops aim to improve the teaching skills of civil engineering faculty. Participants learn what constitutes effective teaching and how to apply learning style models to the organization and conduct of a class, use classroom assessment techniques to assess student learning, organize a class, deliver classroom instruction, assess a class from a student's perspective, and self-assess their classes. Member volunteers develop and refine the workshop curriculum. The six-day hands-on workshop provides seminars, demonstration classes by master teachers, and the opportunity for participants to teach a class and receive feedback from mentors and assistant mentors.

To ensure that the curriculum stays current, the ASCE Department Heads Coordinating Council listens to input from department heads on the needs of their faculty. Workshop faculty are recruited from previous participants, first as assistant mentors and then as full mentors. A mentor and an assistant mentor are provided for every four participants in the workshop, and they give personalized feedback to participants throughout the event. Since 1999 more than 800 people have participated in the workshops, and a number of graduates have gone on to win accolades at their school and in nationwide contests.

For more information visit www.asce.org/exceed.

ASME Graduate Teaching Fellows *(Aisha Lawrey)*

The Graduate Teaching Fellowship Program is a collaboration between the American Society of Mechanical Engineers (ASME) and US mechanical engineering departments to support and encourage outstanding doctoral candidates—particularly women and underrepresented minorities—in mechanical engineering education and related engineering fields. Fellowship awards of $5,000 a year are given for a maximum of two years. Fellows are selected (or renewed) annually by the ASME Board on Education. Applicants must be PhD students in mechanical engineering with a demonstrated interest in an academic career. Fellows are required to teach at least one lecture course.

The program aims to inspire the next generation of mechanical engineering faculty members through society and university collaboration. There have been 58 fellows since 1992, 80 percent of whom are now in academic careers and 51 percent of whom are women. ASME plans to continue to grow the program to increase faculty diversity nationwide.

For more information visit https://www.asme.org/career-education/scholarships-and-grants/scholarship-and-loans/graduate-teaching-fellowships.

AIChE Process-Safety Modules *(Phil Westmoreland)*

Safety and Chemical Engineering Education (SAChE) Process-Safety Modules aim to establish and support manufacturing safety. Developed through the AIChE Academy to meet ABET safety expectations through industry leadership, three major levels of online modules define the program. The first, introduced early in a student's education, addresses the importance of process safety, hazard recognition, identification and minimization of process safety hazards, and management of process safety hazards. The second level focuses on core undergraduate modules: understanding hazards and risks, processing safety at a personal level, managing hazards and risk, and assessing hazards. The third level involves advanced or elective modules, including those for industry use; topics include safe design and operation, equipment hazards, quantitative methods and hazard assessment, risk-based process safety management, and materials hazards. New modules are being developed with the goal of ensuring that all graduating bachelor-degree chemical engineers anywhere in the world are knowledgeable about process safety.

For more information visit http://sache.org/.

IEEE Standards Simulation Workshops *(Larry Larson)*

In 2015 and 2016 Texas State University partnered with the IEEE Standards Education Program to design a one-day workshop demonstrating technical standards development. The interactive workshop includes short introductory lectures and case studies, presented by technical experts and leading standards developers from industry, and a table-top working group simulation to reach consensus on "standards" related to communication, transportation, and power. In the second half of the workshop, each participant is assigned a role modeled on real-life motivations of those participating in standards development.

The workshops are now available as a licensed box game with player materials, videos, and an instructional manual, for use without a professional facilitator. Texas State and IEEE are developing a version of the workshop that can be integrated into coursework as a module rather than as a full-day event and that can be mapped to ABET evaluation criteria.

For more information visit www.ieee.org.

Using Societies to Facilitate Academia-Industry Alignment

ASME Vision 2030 *(Bill Wepfer)*

ASME's Vision 2030 project (V2030) aims to (1) define the knowledge and skills that mechanical engineering and mechanical engineering technology graduates should have to be globally competitive, and (2) advocate for the adoption of recommendations for mechanical engineering curricula with the goal of better preparing graduates to meet the demands of a changing professional environment. To that end the project analyzed the perspectives of over 1,400 engineering managers in industry, more than 1,100 recent mechanical engineering graduates, and mechanical engineering education leaders from 80 universities on how mechanical engineers should be educated to meet the current and future demands of a transforming profession.

Several aspects of the educational landscape emerged as target areas for change, encompassing a range of educational pathways to accommodate the increasingly diverse practice of mechanical engineering. Target areas include richer practice-based experience, stronger professional skills, more flexible curricula, greater innovation and creativity, technical depth specialization, and a new balance of faculty skills. Flexible curricula are especially important in creating a student-elective array of mechanical/multidisciplinary options in programs for majors or minors.

The project also resulted in changes to the ABET mechanical engineering program criteria to support more flexibility and greater emphasis on design through product realization requirements. The program emphasizes "design-make-innovate-create" with a strong element of professional skill development. Nearly half of the departments surveyed have used V2030 to leverage curriculum change and resource acquisition.

For more information visit https://community.asme.org/board_education/w/wiki/7883.asme-vision-2030-project.aspx.

AIChE Chemical Engineering Academia-Industry Alignment *(Jim Hill)*

The AIChE report *Chemical Engineering Academia-Industry Alignment: Expectations about New Graduates* (2015) lays out five objectives: (1) obtain opinions on the preparation of undergraduates and PhD graduates for the jobs they take; (2) determine whether graduates need more workplace preparation, and if so in what areas; (3) assess a number of subject areas in terms of career importance, level of academic preparedness, and the need for more academic preparation; (4) identify areas of growing career opportunities for chemical engineering graduates; and (5) evaluate the need for practical and/or intern experience for chemical engineering undergraduate students, graduate students, and faculty. In addition, the study assesses a possibly problematic shift in faculty expertise away from core areas of chemical engineering.

Key messages of the report include the importance of theory and fundamentals based on physical understanding; broadening of topics (not all with equal depth); instillation of classical knowledge and critical thinking; and development of communication skills, teamwork/leadership skills, and open-ended problem solving. Key conclusions were that there has been a shift in faculty research interests in current chemical engineering programs; practical experience in chemical engineering is valued yet few institutions require it; and academic institutions do not feel there is as much need for workplace preparation as do those in industry. However, the study also suggested that chemical engineering education is more aligned with industry needs than hypothesized.

The study is available at https://www.aiche.org/sites/default/files/docs/conferences/2015che_academicindustryalignmentstudy.compressed.pdf.

Societies and Informal Learning[1]

EWB-USA Project-Based Learning through an International Community Program (Cathy Leslie)

An important part of the Engineers Without Borders (EWB)–USA mission is to provide education and training for the next generation of engineers. The International Community Program creates long-term relationships (typically five years) between student chapters and underserved communities in other countries. These partnerships assess, design, construct, and monitor two or three types of infrastructure projects. Under the guidance of a professional mentor, members develop and use skills in project management and design as well as leadership, communication, time management, persuasion, negotiation, and fund raising.

Students must determine what materials are locally available and what a community can afford now and in the future. They assess and create alternatives, design, raise funds for travel and construction, arrange for the delivery

[1] The presentation on IISE Certificate Training listed in the workshop agenda was not given as the speaker was unable to attend.

of materials, construct, train for operations and maintenance, and monitor and evaluate.

Many of the benefits of participation have been documented through qualitative feedback from graduated students and corporate partners as well as a broad ongoing study sponsored by NSF. The study indicates that EWB participants demonstrate higher than average professional skills from their experience leading teams, managing projects, and working in cross-cultural and cross-disciplinary environments. These skills make them coveted recruits at the world's premier engineering firms and enhance the image and marketability of the universities that host approximately 150 EWB-USA chapters.

For more information visit www.ewb-usa.org.

SAMPE Bridge Contest *(Karin Anderson)*

The Society for the Advancement of Material and Process Engineering (SAMPE) is a worldwide technical society with both professional and student members from academia, government, and industry. Because members work in all areas of material and process development, serving their needs and interests can be challenging. The SAMPE Bridge Contest was launched 20 years ago to address this issue.

Student members are asked to design, analyze, fabricate, and test two-foot-long bridges made of advanced materials and processes. They frequently use engineering software that is standard in the industry, and are asked to present their data in a style similar to presentations in their future careers. They often receive credit for the project through a professor, a senior project, or a course. Industry sponsors are given the opportunity to showcase their materials to prospective engineers through the donation of materials and monetary awards for students. Many of the students also meet with industries in their local communities and ask companies to help them build their bridges, thereby fostering collaboration and student learning.

The contest is held throughout the United States, Canada, Mexico, Brazil, China, and Japan. The winners meet and compete globally, developing both cultural and engineering experience, and new technology categories are continually integrated into the program.

For more student program information visit www.nasampe.org/page/students.

6

From Analysis to Action

Throughout the workshop subgroups of participants met in breakout sessions to discuss specific issues associated with the role of professional societies in US engineering education. In the first two rounds of breakout sessions, participants examined eight topics tied to issues raised in the lightning rounds.

First set of breakout sessions:

- developing partnerships for innovation in education
- promoting diversity
- fostering interdisciplinary engineering education
- raising awareness of engineering disciplines.

Second set of breakout sessions:

- fostering alignment among societies
- fostering societies' alignment with academia
- fostering alignment between academia and industry
- fostering societies' alignment with informal learning.

At the end of the workshop's first day, participants identified a long list of issues raised during the presentations and discussions and then voted on those they wanted to examine in breakout sessions on the second day. The following six sets of questions resulted from this process:

- How can engineering societies share effective practices from the 50K Coalition initiative? How can more societies get involved in the initiative? How can the 50K initiative provide a framework for setting targets for the number of female and underrepresented minority faculty?
- How can engineering societies improve the public perception of engineering via marketing (à la NCAA)?
- What role can engineering societies play in helping engineering education align with the pace of change in the field?
- What role can engineering societies play in influencing the criteria for faculty success, including promotion, tenure, and recognition?
- How can engineering societies undertake joint projects and design competitions using the Grand Challenges framework?
- What is the role of engineering societies in providing training as part of engineering education?

This chapter summarizes the plenary session reports of the breakout group representatives, along with the concluding remarks of the chair of the workshop planning committee and plans for follow-up meetings to build on the progress made at the workshop.

DISCUSSION TOPICS RAISED IN LIGHTNING ROUNDS

Innovation

Two significant themes emerged in the subgroup discussion of collaborations among societies on innovations in engineering education, said Kristine Ward, who reported to the plenary session for this subgroup. The first is that effective collaborations are project based. "Going in with some vague ideas or some themes that you may want to work around usually doesn't get to an effective outcome," she said. "Project-based [initiatives] with a tangible result are usually the best ways to collaborate as societies."

The second theme was the need to rally around larger initiatives in engineering education to effect change. For example, the creation of the $15 million movie "Dream Big," which was spearheaded by the American Society of

Civil Engineers, was designed to inspire young people to dream about what they could accomplish as engineers. Similarly, student competitions are a way to both inspire students and enable societies and academic institutions to collaborate more effectively.

Resources pose both opportunities and constraints, Ward observed. Among resources that offer opportunities, she cited the Frameworks Institute (frameworksinstitute.org), which provides tools for communication, research, and other activities to help convey the value of an education in STEM subjects. For example, talking to K–12 students is very different from talking to college students, which in turn is very different from talking to millennials in the workforce, and advisors can help differentiate among these audiences. Another online resource is the Portal to the Public (https://popnet.pacificsciencecenter.org), run by a collaboration of institutions dedicated to sharing ideas and strategies for scientist-public engagement.

A specific proposal would be to establish an ASEE task force to develop ideas about undergraduate engineering education that could be picked up by disciplinary societies.

Diversity

The siloing of fields in engineering contributes to the challenge of increasing diversity, reported Albert Manero from this subgroup. Funding differences, the balance of risks and rewards, and communication are also barriers to greater diversity to be addressed.

Successful strategies and success stories can affect not only underrepresented groups. An impact on a small group can in turn have an impact on an entire industry, Manero said, especially if that impact is well publicized and communicated.

Partnerships among sectors can help address diversity issues. For example, collaborating horizontally across academia and industry can improve the odds of success for diversity initiatives. Grant agencies can further this process by requiring representation and inclusion, broadening the criteria by which proposals are judged.

People need to participate, contribute, and succeed to feel that they are more than just a representative of a group, Manero pointed out. Having social impact and creating social change are metrics by which members of a team judge themselves and their work. Full integration into a group, rather than being on the fringes, can create greater collective impact and benefit all members of the group.

Interdisciplinarity

The subgroup on fostering interdisciplinary engineering education noted that interdisciplinarity does not mean being an expert in every area, said Burton Dicht. Rather, it means being able to interface and work as an effective team member with people who have different disciplinary backgrounds, even outside of engineering. Some university programs have students working outside their discipline (such as finance majors doing engineering design, or engineering students doing logistics). Many programs fuse different disciplines (bioengineering is a good example), and capstone projects are good tools to drive interdisciplinary work, Dicht observed.

Other types of programs and actions could enhance interdisciplinary engineering education, Dicht continued, by, for example, linking technical professional societies to the engineering curriculum, giving students credit for work they do with professional societies, and defining learning objectives and assessments that incorporate interdisciplinary objectives. These activities could be linked with ABET criteria and include incentives for faculty (such as teaching credits) to drive the activities.

Starting these activities with freshmen would introduce them early on to interdisciplinary education. In addition, such an approach could attract students from other disciplines like mathematics or the sciences.

However, efforts to increase interdisciplinary engineering education also face a number of obstacles, Dicht noted, such as cultural issues within departments, lack of credit for faculty in pursuing such objectives, already high teaching loads, and silos between disciplines.

One innovative approach discussed by the subgroup would be to create a multidisciplinary challenge among engineering societies. For example, societies could collaborate to create a competition or challenge that would go to all students and require multidisciplinary teams as a condition of entry.

Awareness

Engineers do not do a good job of marketing their profession, said Charles Reinholtz, who reported from the subgroup that discussed raising awareness of engineering disciplines. From high school students to the general public, people are not much aware of what engineers do and why it is exciting. In contrast, he said, look at the National Collegiate Athletic Association (NCAA) and college athletics. National signing day is a well-covered event that generates interest and excitement among large segments of the population.

Why can't the same thing be done for engineering, Reinholtz asked? Generating excitement about high school and college-level engineering could raise media and public awareness of the profession. Competitions, the Grand Challenges for Engineering, scholarships, goal-oriented teams, and a supportive culture could all get students and others excited about engineering.

Furthermore, Reinholtz pointed out, engineering has something that the NCAA does not: jobs after graduation. Just a tiny fraction of student athletes go on to become professional athletes, whereas engineers have tremendous job prospects after graduation. This career potential could drive a much greater effort at the high school and college level to promote engineering.

The group also discussed the gap that often emerges between students' membership in professional societies and their involvement in those societies as professionals. In the first several years after college, beginning professionals tend to lose interest in societies and rejoin only later. Reduced or free membership for engineers for the first year or few years after they graduate could close this gap, Reinholtz suggested.

Engineering and the products of engineering also could be humanized to a greater extent. A focus on individuals, groups, and cooperation can form the basis of success stories that could interest students and others in engineering. For example, stories of engineers from disadvantaged circumstances who succeed can inspire young people to enter the profession.

Finally, it is important to get media and marketing people involved in communicating about engineering, Reinholtz said. Competitions, success stories, and societal impact can all get the media interested. He suggested emulating NASCAR, which continually tweaks the rules of auto racing to make its events more exciting. Competitions that are too long or poorly formatted are less likely to excite students and the public, even though engineering has unique stories to tell.

Fostering Alignment among Societies

The subgroup on establishing effective intersociety collaborations identified several roadblocks to such collaboration, Burton Dicht reported, including a constantly changing roster of volunteers, uneven distribution of workload, difficulty finding the right contacts at other societies, challenges identifying subject matter experts to address key issues, intellectual property held by societies, and competition for the same audiences and funding. The sustainability of collaborations is also an issue, since societies sometimes lose interest in a topic.

The subgroup discussed several ideas intended to overcome the roadblocks. One is to have more workshops that bring societies together to discuss topics of common interest. For example, ABET brings together societies to address issues, and they are not always specifically associated with accreditation. Societies also have to be willing to explore topics beyond their own fields of interest, Dicht remarked, including broader issues such as the Grand Challenges. Areas such as curriculum development could provide a further basis for collaborations.

Societies have areas of strength and weakness, but they can all focus on broad objectives. And new social media tools can connect people, whether specifically with a society or more broadly.

Reciprocal memberships in societies bring together the staffs of those societies to learn about how other organizations operate. Transparency and trust among societies foster the sharing of both good and bad practices. Research on success stories and dissemination of those stories and lessons learned can encourage other societies to try something new. Societies need to play the long game, said Dicht, and not try to do too much. Trying things on a small scale with specific endpoints can lead to bigger efforts and help build relationships.

The conversation started at the workshop needs to continue, Dicht said, whether face to face or virtually. A central clearinghouse supported by a robust infrastructure could store, connect, and disseminate ideas so that information is easier to access than it is today.

Fostering Alignment with Academia

Engineering societies incorporate both academic and industrial members and cultures. As such, they are particularly well positioned to help universities prepare students as professionals, said Phillip Westmoreland, who reported for the subgroup on interactions between societies and academia. Professional societies shape expectations about professional behaviors in both technical and nontechnical areas, and they can analyze and shape curricula, thereby helping determine the future of professions.

Some of the influence of societies is exerted through student memberships and student chapters, which are excellent opportunities for building leadership and other skills, said Westmoreland. Chapter advisors can serve as direct connections to societies and also provide continuity as students cycle through their educational years. Affinity and disciplinary societies can support each other's efforts both with students and in academia through work on both technical topics and pedagogical approaches.

Societies also have a subtler role to play in presenting the results of academic research to the public. Many people have faith in the role of engineers as honest brokers, Westmoreland observed, even if they do not necessarily trust the companies with which engineers are associated. Societies can advance understanding of uncertainties, interpretation of research results, and applications of those results. They can inform the public, help regulators make better decisions, and encourage other valuable uses of new knowledge. An example, said Westmoreland, is helping the public understand the balance between cyberpower and cyberthreats. In these ways, societies can enable action despite uncertainty.

Fostering Alignment between Academia and Industry

Societies can enhance alignment between academia and industry, reported Harriet Nembhard, through, for example, competitions that result in commercial products, cooperative agreements, and internships. Industry speakers and professors associated with industry can work with students and faculty members to disseminate the perspectives of industry.

Faculty members can engage with industry through advisory boards, and societies can facilitate these links. Other ways to connect faculty with industry are through partial industrial funding of faculty startup packages or through summer internships or immersion experiences in industry. NSF-sponsored Industry-University Cooperative Research Centers already have the charge of increasing the alignment between faculty members and industry, and their missions could be broadened to other kinds of collaborative activities.

The academic culture is not always supportive of links with industry, Nembhard observed. Changes in how promotion and tenure are assessed could make faculty members more open to working with industry and contribute to broader changes in the academic culture.

Fostering Alignment with Informal Learning

Informal learning outside of a structured classroom is often a major part of the college experience. In fact, research suggests that much of the learning retained from college comes from informal activities, reported Albert Manero.

Informal learning can be particularly useful in attracting and retaining groups that are underrepresented in STEM subjects, such as underrepresented minorities and women. Societies can help attract the members of these groups in several ways. They can help with transportation, families, time demands, and other needed forms of support. They can offer in-person or virtual mentoring that helps engage students and provides them with a sense of identity and belonging. They can help students overcome bias that discourages their participation in such programs.

Efforts to increase diversity and inclusivity can create a feedback loop that accelerates the process of broadening participation, Manero observed, to ensure that future cohorts of students, faculty members, and mentors are more effective and deeply engaged.

DISCUSSION TOPICS CHOSEN BY PARTICIPANTS

The 50K Coalition

The 50K Coalition, although still relatively new, is doing a good job of sharing effective practices, reported Leah Jamieson on behalf of this breakout group. It is holding bimonthly webinars, having meetings to share best practices around six common themes, and developing online dashboards. At the time of the workshop, the coalition had 11 university members and was working on scaling up. Jamieson pointed out that increases in size may give rise to issues of resources and data reconciliation that have to be resolved.

One promising approach is the development of models for "mutually collective impact" that rely on common goals, including a framework for contributing to these goals outside a membership model. Getting the word out about the coalition is also important, so that other colleges and universities and engineering societies become interested in contributing to its efforts. Distribution of a white paper about the coalition to societies could foster internal discussions about participating so that societies understand how the goals of the coalition mesh with their goals.

The breakout group discussed the possibility of setting targets for the numbers of female and underrepresented minority faculty members, heads of departments, and deans. (Although not an integral part of the coalition's efforts, it could be a parallel effort.) The relevant organizations could connect with each other to lay out a process for setting bold and realistic targets, including ownership of the targets and the tracking of progress. The presence of more female and underrepresented minorities could drive efforts to diversify the faculty and vice versa, Jamieson pointed out.

Improving the Public Perception of Engineering

Engineering departments at colleges and universities and engineering societies do not do a good job of publicizing the things they do, said Gregory Washington, reporting for the subgroup that discussed improving the public perception of engineering. One way to improve public perceptions of engineering would be to create a depository for information that highlights what is being done by these entities, both at individual institutions and across institutions.

The development of a set of "big ideas" that teams of engineers could work on also could promote the field. As examples Washington mentioned the development of autonomous or driverless vehicles, space launches, and robots that incorporate artificial intelligence. If engineering societies oversaw these projects, they could hold conferences on the topics and promote them both regionally and nationally.

The subgroup proposed partnerships with experts in marketing. Many engineering departments and societies are engaged in interesting projects; marketing assistance could help these projects gain broader visibility, which could promote the discipline as a whole.

Finally, the group called attention to the need to develop a coherent and broadly applicable definition of engineering. "It's hard to define what it is we do because we give [people] so many different definitions," Washington said. A single definition that extends across disciplines—such as Dan Mote's definition of engineering as "creating solutions serving the welfare of humanity and the needs of society"—would establish a baseline from which individual organizations can develop more specific definitions of what they do.

Aligning Engineering Education with the Pace of Change in the Field

Engineering societies could be much more directly involved with translating information from industry into the curriculum, reported Kodi Verhalen. One way would be through a virtual society fair modeled on the virtual career fairs that companies have begun to hold for engineering students. In a virtual fair, individuals can participate wherever they are rather than coming to a central location to meet with representatives of organizations. Through virtual society fairs, faculty members and students could learn what societies are doing and what educational resources are available. In addition, students could learn what societies exist in their fields and how those societies can help them as they graduate and move into professions.

Engineering educators also have a responsibility to inform their students about the opportunities societies provide. Many students learn about technical societies but do not necessarily learn about the professional societies that can help them develop their practice, continue learning, and prepare for changes in their profession.

A central clearinghouse could help bridge the gap between changes in industry and changes in engineering education. For example, a web-based system targeted to faculty and students could provide information on design competitions, scholarships, prizes, openings to serve as mentors for K–12 students, and other opportunities. A clearinghouse could convey educational materials from technical and professional societies to faculty and students, such as webinars on what an organization does, educational modules on specific topic areas, or material on the pace of change in a technology. It also could inform faculty members and students about conferences, including those specifically targeted at students.

Finally, Verhalen pointed out that although some defined engineering bodies of knowledge exist, they are updated infrequently, which means that they cannot necessarily keep pace with changes in engineering and industry.

Influencing the Criteria for Faculty Success

One important step that could influence the criteria for faculty success is gathering more information about nontenured faculty, said subgroup reporter Anastasios Lyrintzis. The task would not be difficult and could provide useful information, especially about women and underrepresented minorities in this group.

The subgroup also discussed the possibilities for societies to determine the impact of publications, which are critical to promotion and tenure. Societies could provide guidance on how best to use them for this purpose.

Workshops for department chairs and deans also could generate awareness of how best to help assistant and associate professors, such as through mentoring. The efforts of the American Society of Mechanical Engineers could serve as a model in this regard, Lyrintzis suggested.

Promotion and tenure could be discussed at the annual American Society for Engineering Education meeting of deans. This or other meetings also could be broadened to include other people involved in these matters.

A particular issue that needs to be explored is the role of nontenure-track faculty, who are sometimes treated as second-class citizens. For example, could titles be changed so that people are recognized by their area of expertise, such as teaching or practice, rather than their position in an organizational hierarchy? Could metrics be developed to measure the impact of different kinds of activities? Getting full professors to adopt such measures could be difficult, Lyrintzis acknowledged, but good arguments can change minds.

Design Competitions Using the Grand Challenges Framework

The Grand Challenges for Engineering offer tremendous opportunities to engage students at all levels in engineering problems of major importance to society. Drawbacks to this approach include the bandwidth of society staff and volunteers, engaging industry in the effort, and the proper role of societies, said Burton Dicht in his summary of the subgroup discussion. Societies need to be enlisted at the top level, with the involvement of staff, volunteers, and other stakeholders, including faculty members and students.

One question is whether existing competitions can be used as models for such an effort, repackaged to incorporate the Grand Challenges. What about involving other disciplinary associations such as those in law, business, or medicine, since these professionals also would be involved in solving the Grand Challenges? Finally, could such a competition be aligned with student capstone projects?

The American Association of Engineering Societies has agreed to serve as a facilitator to bring societies together, to curate existing society competitions, and to create a database that could foster buy-in and planning, and many other societies have expressed initial interest in this approach, Dicht reported. These societies could create an outline for a joint competition proposal describing how it would work, the competencies required to move forward, and a communication plan to involve other societies. An industry outreach plan could involve businesses.

Important goals are to finalize agreement on which societies will take part, define the competition framework, define society roles and responsibilities, develop a competition timeline with target dates, and develop a marketing and communications plan. Dicht cited the July 2017 Grand Challenges Summit, to be held in Washington, DC, as an opportunity to further the plan.

The Role of Societies in Providing Training

One way for societies to become more involved in providing training to faculty members and other educators would be for ABET to convene societies to develop workshops for faculty members on teaching engineering, reported Anne Spence. Such workshops could convey information both from the societies to educators (for example, about the skills and content knowledge that graduates lack) and in the opposite direction, so that societies learn more about what educators need.

Engineering instruction in colleges and universities is also related to the preparation of K–12 teachers of engineering. Faculty members and departments could be identified to serve as advocates for teacher preparation and to forge strong connections with societies. Together, these individuals and societies could work to develop and disseminate webinars, podcasts, and other tools that move K–12 engineering education forward.

CONCLUDING REMARKS

The workshop marked the beginning of a process, not its culmination, said Leah Jamieson, noting that the project will continue with follow-up meetings. She encouraged the participants to "Keep talking, because what got accomplished here and the things that we're all hoping will come out of it are happening because people are connecting and talking to each other, and that's at the heart of getting started. It doesn't say where we're going to end up, but it certainly is essential to getting started."

After the workshop the steering committee and sponsor decided to focus five follow-up meetings on the topics identified by participants for discussion on the workshop's second day. Each meeting will cover one of the topics.[1] These meetings will allow for examination of the issues identified at the workshop in greater detail, with proceedings published separately.

[1] The topics of "Aligning Engineering Education with the Pace of Change in the Field" and "The Role of Societies in Providing Training" will be merged.

Appendixes

Appendix A

Workshop Agenda

THURSDAY, JANUARY 26, 2017

08:30 AM – 09:00 AM
Registration (Continental Breakfast available)

09:00 AM – 09:30 AM (30 min.)
Welcome, Remarks from the Sponsor, and Goals for the Day
Alton Romig, Jr., Executive Officer, National Academy of Engineering
Leah Jamieson, Chair, Workshop Steering Committee
Elliot Douglas, National Science Foundation

09:30 AM – 10:00 AM (30 min.)
Stage-setting Presentation: Conclusions from survey and interviews, outline of Big Questions
Leah Jamieson, Chair, Workshop Steering Committee

10:00 AM – 10:30 AM (30 min.)
Lightning Round #1: examples/case studies on "what we do"

10:30 AM – 10:40 AM (10 min.)
Break (and move to breakout sessions)

10:40 AM – 11:25 AM (45 min.)
Breakout #1 – discussions about topics raised in the case study and examples (clustered into similar activities)

11:25 AM – 11:45 AM (20 min.)
Breakout Reports

11:45 AM – 12:30 PM (45 min.)
Plenary – issues of assessment
Barbara Bogue, PI and cofounder, SWE AWE (Society of Women Engineers' Assessing Women and Men in Engineering) Project
Betty Shanahan, Co-PI, SWE AWE

12:30 PM – 1:45 PM (75 min.)
Lunch and Networking (lunch provided)

1:45 PM – 2:15 PM (30 min.)
Lightning Round #2: issues/problems/barriers faced and overcome (how we do it)

2:15 PM – 3:30 PM (45 min.)
Breakout #2 – discussions about topics raised on issues/barriers (clustered into similar activities)

3:30 PM – 3:45 PM (15 min.)
Break

3:45 PM – 4:15 PM (30 min.)
Breakout Reports

4:15 PM – 5:15 PM (60 min.)
Collaboration Sessions Planning Town Hall
Leah Jamieson, Chair, Workshop Steering Committee
Ken Jarboe, National Academy of Engineering

5:15 PM – 5:30 PM (15 min.)
Final Comments
Leah Jamieson, Chair, Workshop Steering Committee

5:30 PM
Adjourn

FRIDAY, JANUARY 27, 2017

08:00 AM – 08:30 AM
Registration (Continental Breakfast available)

08:30 AM – 09:15 AM (45 min.)
Keynote – State of Engineering Education
C. D. Mote, Jr. President, National Academy of Engineering
Darryll J. Pines, Dean and Nariman Farvardin Professor of Aerospace Engineering
Clark School of Engineering
University of Maryland, College Park

09:15 AM – 09:20 AM (5 min.)
Reflections from Day 1: Topics for Collaboration Sessions Announced
Leah Jamieson, Chair, Workshop Steering Committee
Kenan Jarboe, Project Director, National Academy of Engineering

09:20 AM – 10:15 AM (55 min.)
Collaboration Breakout Session 1

10:15 AM – 10:30 AM (15 min.)
Break and move to Collaboration Session 2

10:30 AM – 11:30 AM (60 min.)
Collaboration Breakout Session 2

11:30 AM – 12:15 PM (45 min.)
Brief Reports on Collaborations

12:15 PM – 12:30 PM (15 min.)
Final Reflections and Next Steps
Leah Jamieson, Chair, Workshop Steering Committee

12:30 PM
Adjourn and Networking Lunch (lunch provided)

Presentation and Breakout Sessions

What we do topics – Morning

Developing partnerships for innovation in education (student chapters and beyond)

1. Kris Ward: SME Certified Manufacturing Technologist (CMfgT)
2. Randi Rosebluth: SWE Collegiate Leadership Institute
3. Bill Wepfer: ASME Industry Standards Infusion in all four years of ME/MET degree programs
4. Phil Westmoreland: AIChE Manufacturing Institute

Promoting diversity

1. Karl Reid: NSBE Retention Toolkit
2. Glenda La Rue: WEPAN: TECAID – Transforming Engineering Cultures to Advance Inclusion and Diversity

Fostering interdisciplinary engineering education

1. Melissa Prelewicz: AAES Engineering Competencies
2. Libby Jones: EWB-USA Global Classroom

Raising awareness of engineering disciplines

1. Nichol Campana: ASM International "Materials Camp"
2. Bob Fine: ANS Student Conferences
3. Chris Ciuca: SAE Collegiate Design Series

How we do it topics – Afternoon

Establishing effective intersociety collaborations

1. Constance Thompson: NSBE 50K Coalition
2. Bill Mahoney: ASM International "Material Advantage"

Bolstering society-university collaboration

1. Kris Ward: SME E-learning
2. Leslie Nolan: ASCE ExCEEd Teaching Workshop
3. Aisha Lawrey: ASME Graduate Teaching Fellowships
4. Phil Westmoreland: AIChE Chemical Process Safety Curriculum Module
5. Larry Larson: IEEE Standards Organization Student Simulations

Using societies to facilitate academia-industry alignment

1. Bill Wepfer: ASME Vision 2030 Advocacy for Mechanical Engineering Education
2. Jim Hill: AIChE Industry-Academic Alignment Workshop

Societies and informal learning

1. Cathy Leslie: EWB-USA International Community Program
2. Karin Anderson: SAMPE Advance Materials (and Additive Manufacturing) Bridge Building Contest
3. [James Moore: IISE Certificate Training – cancelled]

BREAKOUT SESSIONS: FRIDAY, JANUARY 27, 2017

Round One

How can engineering societies share effective practices from the 50K Coalition initiative? How can more societies get involved in the initiative? How can the 50K initiative provide a framework for setting targets for the number of female and underrepresented minority faculty?
Facilitator: Leah Jamieson
Room: 106

How can engineering societies improve the public perception of engineering via marketing à la NCAA?
Facilitator: Gregory Washington
Room: 103

What role can engineering societies play in helping engineering education align with the pace of change in the field?
Facilitator: John Wall
Room: 105

Round Two

What role can engineering societies play in influencing the criteria for faculty success, including promotion, tenure, and recognition?
Facilitator: Don Giddens
Room: 106

How can engineering societies undertake joint projects and design competitions using the Grand Challenges framework?
Facilitator: Asad Madni
Room: 103

What is the role of engineering societies in providing training as part of engineering education?
Facilitator: Anne Spence
Room: 105

Appendix B

Survey and Interviews

OVERVIEW OF SURVEY AND INTERVIEW FINDINGS

This document highlights some themes that emerged looking across both the survey and interview findings. [Note that the interviews were meant to both provide examples of some of the activities and collaborations of engineering societies, as well as to gather additional information not addressed on the survey (e.g. evaluation strategies) to round out the picture of societies' education efforts].

OVERALL THEMES

All of the societies are primarily concerned with the **professional development** and **continuing education of their membership**. There is also a commitment (particularly among discipline-focused societies) to **elevating the status of engineering and ensuring its future**.

Engineering education at the undergraduate level is not a priority area of focus for all societies, but it ranks highly. Student chapters represent the most common strategy for connecting to higher

The survey was conducted by Inverness Research, which analyzed the results and prepared this report.

education. Pre-college education, while important to many, is not a priority—whether that is due to a lack of resources, or know-how, is not clear. However, those not involved in the pre-college engineering education believe it is critical to ensuring quality candidates for the engineering pipeline.

Nearly all interviewees expressed an interest in the results of the study and learning more about education efforts across the field.

Areas Where Societies Report Being Strong in Their Education Efforts

- Disseminating practices within their own membership through conferences, meetings, newsletters, etc.
- Strengthening the field through professional development and continuing education of their membership
- Creating greater awareness of engineering (or their particular discipline within engineering) through student chapters and other outreach efforts
- Making the case for the importance/relevance of their discipline
- Reaching out to and supporting groups or populations that are underrepresented in engineering
- Partnering with outside organizations or other societies
- Providing certification (discipline-based societies)

Areas Where Societies Could Strengthen Their Education Efforts

- Tracking the details of their education efforts (e.g. budget, investment, human capacity, etc.)
- Evaluating the impact of their programs in more coherent and rigorous ways (counting participants is the most common approach)
- Engaging in precollege engineering education efforts (not all societies are interested in this but they all believe it is important)

- Disseminating their engineering education efforts (interviews revealed dissemination is limited to traditional channels within societies' membership)
- Connecting across societies (there are some examples of strong connections, but there are several barriers as well)

Possible Topics for Discussion at the Workshop

- What are some efficient strategies for societies to engage in collaborative education efforts, or how can societies collaborate and communicate efficiently?
- What are some alternative dissemination strategies, so that promising practices and innovative programs and examples can be shared across the field?
- What types of evaluation approaches make sense for societies to adopt? How can evaluation be designed so that programs improve and others can learn from the outcomes?
- What are some ways societies can address communication barriers across societies? Are meetings and conferences the best way? Are there others?
- How can societies engage faculty and encourage them to change their practice, or introduce innovation into engineering education at the undergraduate level?
- Is there a way to go beyond student chapters as a way to address undergraduate engineering education?
- For those that are interested, and have the financial and human capacity, how can societies support more and better engineering education at the pre-college level?

NAE ENGINEERING SOCIETIES STUDY – SURVEY RESULTS

This document provides highlights of the NAE Engineering Societies survey data collected by Inverness Research in the spring and summer of 2016.

Method

After developing and implementing a pilot survey that included 10 societies, Inverness Research drafted a final survey that was reviewed and approved by the project committee. The NAE and IR then drafted an email invitation that was sent to 121 societies, in some cases to multiple people at a given society. The email asked for societies' participation in the study, and in particular, for recipients to decide who was the best person to complete the survey for their society.

After NAE sent two reminder emails, a total of 58 surveys were completed, for a return rate of 48%. There were eight cases in which two surveys were completed for a single society. In these cases, we decided to either contact the individuals and ask which survey to include in the study, or opted to include the survey of the individual who also participated in an interview, if applicable.

The Sample

The surveys were completed by a range of leaders within the organization, with the most prevalent position being the Executive Director or the President. Respondents with other roles included:

- Board of Directors member
- Director (various areas, such as education, outreach, etc.)
- General manager
- Vice President

The breakdown of responding societies, according to number of members was:

less than 1,000 (small)	10	17%
1,000 - 9,999 (medium)	15	26%
10,000 - 49,999 (large)	21	36%
more than 50,000 (extra large)	12	21%

Because there was a relatively even spread of societies in these initial size groupings, we conducted sub-analyses of the survey questions based on the size bands above. We also conducted sub-analyses according to whether a society was discipline-focused, or an affinity society. In the following summary, we indicate where there were statistically significant differences between these groups.

After an initial review of the survey findings, we conducted an additional sub-analysis to explore the relationship of societies' priority education goals, activities, and audiences, and their rating of education as a priority weighed against other society priorities.

General Findings

- Overall, engineering (or engineering-related) societies are engaged in a range of education activities that target a range of audiences. Nearly every category of goal, activity type, and audience is represented in the work of the societies who responded to our survey.
- Supporting and growing membership is a high priority goal for all of the societies, which means that activities that both increase membership numbers and contribute to the professional growth of (practicing) members are important. Societies are less concerned with influencing policies related to engineering education, or in addressing pre-college engineering (for the most part). A sub-analysis revealed that discipline-based societies are more apt to focus on improving curricula and materials than affinity societies. Further, affinity societies are more apt to focus on culture change than discipline-focused societies are.
- The sub-analysis by membership size did not reveal many significant differences. One notable exception is level of investment in education endeavors, where small- and medium-sized societies are more apt to say their level of investment has stayed the same, while larger societies say their funding for education has increased in the past two years.
- The majority (87%) of societies face some kind of barrier in their engineering education work. The most common barriers include: communication; improving engineering curricula; incentives; as well as issues related to time, resources, and

funding. A sub-analysis revealed that affinity societies are more apt to report facing barriers than discipline- focused societies.

- About ¾ of all respondents are engaged in partnerships with outside organizations, and 86% use connections to at least some extent with other engineering societies. Over half believe these connections to other engineering societies are useful to a good or great extent. There is a wide range of organizations and societies that are engaged in these partnerships.
- Societies' investment in engineering education has not decreased in the last two years. Annual budgets, industry, and university-based faculty are counted as resources for engineering education work for most societies. Further, 77% of respondents said that engineering education is just as or more important than other society priorities.
- A majority (85%) of societies consider themselves leaders in the field. However, half of the societies rated their overall capacity to plan and implement education work as either low or some. 38% rated their capacity as high, and 12% rated it very high.
- Leadership Development is a higher priority for participants who said engineering education was "more or much more important" in the scope of their society's goals and activities compared to those who said engineering education was "less or much less important." The same holds for Continuing Education and Engineering Education Issues/Trends Research - these activities are high priority for those who said engineering education is more or much more important than other society activities. There were no statistically significant differences for target audiences.

Education Goals

Over half of all societies count supporting professional development, leadership development, and increasing diversity as high priority goals. Professional development leads with 90% of societies reporting it as a high priority. Fostering policy changes, and improving curricula and materials are lower priority goals.

A sub-analysis revealed differences between discipline-focused societies and affinity societies. Discipline-focused societies are more apt to focus on improving curricula and materials, even though it is a lower priority overall. Further, affinity societies see culture change as a higher priority than disciplinary societies. There were no significant differences based on society membership size.

In our analysis of the relationship between goals and commitment to education, we found that societies who rated education as important to their society were more likely to identify leadership development as a priority goal.

Education Activities

The majority of all societies are involved in professional development for their membership (82%). Sixty-five percent have student chapters, and 61% provide continuing education. Over half (58%) are working on partnerships with industry, and 51% are concerned with women in engineering. Strong or medium priority activities include mentoring and academic partnerships. Low priority activities include programs for veterans, fellowships, employability training, and student competitions.

A sub-analysis revealed that affinity societies' priority goals are more apt to include fundraising, programs to promote diversity, and pre-college engineering education than discipline-focused societies. However, discipline-focused societies are more apt to include certification as a high priority goal. There were no significant differences in the analysis by society membership size.

In our analysis of the relationship between education activities and commitment to education, we found that societies who rated education as important to their society were more likely to identify Continuing Education and Engineering Education Issues/Trends Research as priority activities.

Target Audiences for Education

Undergraduate students (63%), graduate students (57%), industry (68%), and government agencies (56%) are high priority audiences for over half of societies' education efforts. University faculty (37%) and high school students (33%) are a close second. Low priority audiences include pre-K teachers and students, elementary school teachers and students, middle school teachers and students, and deans/department chairs. There were no significant differences by size, and there were no significant relationships between their rating of the importance of education and their target audiences.

Partnerships

Just over three quarters (77%) of societies said they are engaged in some kind of partnership with an outside agency or organization for their education work.

Professional and technical societies, academic organizations/institutions, government agencies, STEM organizations, and industry were the most commonly cited. At least one society also partnered with the following kinds of organizations:

- Diversity organizations
- International development organizations
- Engineering education organizations
- Accrediting bodies
- Private organizations
- Humanitarian organizations
- Consulting Engineers
- Museums
- Manufacturing Institute
- Mentoring organizations
- Girl-serving organizations
- Non-profit
- Media outlets
- Other standards developing organizations (SDOs)
- State affiliates
- Company that focuses on webinars and other distance-learning for environmental issues and engineering geology

General Program Information

Not all respondents answered all of the following questions. We indicate the number for each question.

Numbers served (N = 54)

41% of respondents did not know how many people were served by their education efforts. Of those who answered, the majority (43%) serves up to 10,000 people.

Numbers of volunteers participating in education work (N=54)

Just under 1/3 did not know how many volunteers participate in education work. Of those who answered, the 50% have up to 500 volunteers.

Annual budget for education (N=50)

34% (17) did not know the annual budget for education. Of those who answered, there was a range of numbers, indicated below:

less than $10,000	2	4%
$10,000 to $99,999	7	14%
$100,000 - $999,999	11	22%
$1MM to $10MM	12	24%
over $10MM	1	2%

Student Chapters (N=41)

Just under ½ of those responding to this question have less than 50 student chapters. Most of the remaining societies have over 50, upwards of 200. Seven societies have over 200 student chapters.

Program administration (N=51)

Societies' central offices administer around 75% of programs for just under ½ of those who answered this question. Local chapters and/or divisions administer up to 50% of programs for just over half of the societies. Student chapters administer up to 25% of programs for just under half of societies. For 33% of societies, student chapters do not administer any of their education programs.

Level of investment in engineering education

The level of investment in education for societies has either increased (51%) or stayed the same (44%) for most societies. It has decreased for 4%, and 2% did not know.

A sub-analysis of the data by membership size revealed a significant difference with respect to recent change in level of investment in engineering education. Small and medium societies' level of investment in engineering education has stayed the same, whereas larger societies have increased their funding for education in the last two years.

Capacity for engineering education

Societies' capacity to plan and implement education work is essentially split—half of the societies rated their overall capacity to plan and implement education work as either low or some. Forty-one percent rated their capacity high, and 9% rated it very high. There were no differences based on membership size.

Resources for Education

The large majority of support for education comes from societies' annual operating budget (93%). Following that, resources include: membership-industry (70%), corporate sponsorship (72%), membership-faculty (67%), membership—college/academic department leaders (52%). Student members (46%), foundations (41%), internal research and/or evaluation results (30%), literature (24%) and NSF funding (24%) are also used. NAE and ASEE publications were resources for just 20 societies of the total sample. There were no differences based on membership size.

Connections with Other societies

Eighty-seven percent of respondent use connections with other engineering societies or organizations at least a little. Eleven percent (6 societies) said they use them "a lot." Of those who use them at least a little, 43% believe that these connections are currently beneficial to some extent. Thirty-four percent believe they are useful to a good extent, and 23% to a great extent. All of those who do not currently use connections to other societies believe they could be beneficial to at least some extent. There were no differences based on membership size.

Barriers

Eighty-seven percent of societies face some kind of barrier in their education work to at least some extent. The majority of examples of barriers described by respondents fell into the following categories:

- communication issues (e.g., connecting members to educators; meetings)
- curriculum related issues (e.g., challenges with changing accepted curricula)
- incentive issues (e.g., getting faculty to change their practice)
- time, resource, funding issues (e.g., lack of resources to be able to scale local programs into a repeatable framework for national level use)

Other barriers were more specific to the discipline or particular mission, such as needing background checks for members to work in schools, lack of identity of a specialty, finding the right partners, improving student access, finding speakers, and the like.

Sub-analysis revealed that affinity societies are more apt to face barriers to their education efforts than discipline-focused societies. There were no significant differences based on membership size.

Leadership in Education

Eighty-five percent of societies consider themselves leaders in engineering education to at least some extent, and 17% of those to a great extent. For many of them, this refers to their particular specialty (e.g. conservation engineering). With respect to the relative importance of engineering education in their society, 78% of respondents believe that education is at least as important than other goals or activities pursued by their society, with 36% saying it is more or much more important. Only 23% (12 societies) said education is much less or less important.

NAE ENGINEERING SOCIETIES STUDY— INTERVIEW RESULTS

The following is a high-level summary of the 30 interviews of engineering societies conducted by Inverness Research. Most of the individuals we interviewed were Executive Directors or Presidents, although there were a few who held other leadership positions. It is worth noting that many of the interviewees began by saying that they did not see their society as an "engineering society," per se, but one that influences and is influenced by engineering, and includes engineers in the membership.

Goals and Activities

Professional development for practicing engineers is a common focus for societies. Supporting and growing membership is also a goal. Across the sample, it is evident that societies engage in a range of activities depending on the needs of their membership, their available expertise and capacity, and funding. Further, societies may organize geographically and provide different kinds of opportunities for different, local audiences.

For the smaller or specialty societies, a common priority goal is to spread awareness of their particular branch of engineering. They also have some expertise in outreach to pre-college and undergraduates to educate engineers and others about their particular specialty.

Other goals and activities that interviewees mentioned include: engaging students in the community and service learning; ongoing professional development for practicing engineers; ensuring the next generation of engineers is prepared to practice engineering; expanding and protecting the reputation of engineering; professionalizing the engineering field; providing a forum for engineers to interact; promoting quality engineering education through ABET; facilitating career transitions; providing research-based design standards; working with regulators, legislators, and policy makers; and providing more application-based programs.

Concerns about the extent to which students graduating from engineering programs are ill-prepared to work in industry were shared by several societies, as exemplified by this statement:

The biggest complaint we get from our membership of 56,000 engineers is that engineers graduating from engineering programs are not well equipped on an application basis to participate in our industry. There are lots of reasons behind that. So we feel a duty to pick up that mantel and really focus on teaching application, how do you actually do what these folks are asked to do in our industry?

Also important for many societies is retaining and supporting diversity in engineering. One interviewee recalled her own experience as an engineering student to explain how important her society's goals are for many:

> I was [an engineering] student in the 70s. It was [this society] that kept me from quitting. Back then there was a lot going on that confirmed I didn't belong. Having a safe place was important—a place where I could exhale and talk to someone who could relate.

Another interviewee described their goals as:

> Anything that touches on the precollege, undergrad, lifelong learning of an engineer . . . we want to ensure it is available, accessible, and excellent.

Activities of societies include: ensuring that degree programs are preparing students for engineering jobs/practice; creating awareness of the discipline (e.g., environmental engineering); providing mentor programs for high school and undergraduate students; providing education for certification programs; providing internship programs to support transitions to work; providing curriculum or support for curriculum at the undergraduate level; providing scholarships; creating and offering webinars and workshops for continuing education; offering accreditation through ABET; offering an early career faculty program; endorsing existing programs such as FIRST Robotics and Project Lead The Way; offering courses and workshops at annual meetings; and partnering students with practicing professionals.

One large society is addressing both the faculty and student experience: it has begun a program for early career faculty to provide them with resources (2-4 hour workshops, networking, mentoring) that will better prepare them for teaching at the university level, and a student program that helps them learn the "difficult to learn" subject matter.

A few societies are more focused on the technician—education programs aimed at high school and community college students who would not likely complete a 4-year engineering degree but aim to work in a job providing engineering or technical support to the engineer.

Overall, most activities of societies are designed to meet the needs of the membership. There are a couple of societies with longer-range views. Finally, there is a growing emphasis on providing programming virtually—through online courses, webinars, and the like.

Evaluation

Most of the societies attempt to collect feedback from participants in their most significant programs but most interviewees noted that they would like to do more to evaluate their work. They tend to collect numbers of participants and programs as indicators of success. As one interviewee said, their evaluation is:

> Almost entirely by numbers: numbers of students who receive fellowships or scholarships, number of dollars that go into the fellowships and scholarships, number of individuals or organizations that contribute to the association, number of active chapters.

Many also conduct member satisfaction surveys (i.e. people vote with their feet and wallets), or surveys that help them understand the professional development needs of their membership.

Some interviewees mentioned wishing that they did more to evaluate the longer-term impact of their work. One interviewee said, "*Everyone struggles with that and it takes lots of resources to figure out what are the right metrics.*"

DISSEMINATION

Most societies disseminate their work, education or otherwise, through annual meetings/conferences, journals, websites, member newsletters, and the like. One interviewee said that a formal venue for disseminating or sharing work related to undergraduate engineering education does not exist, but should.

CONNECTIONS WITH OTHER SOCIETIES

Most societies have at least some connections with other engineering societies, although the connections may or may not involve their work in education. Many invite other societies to their meetings, or attend others societies' meetings.

However, few have made substantial connections that have resulted in collaborative projects. Mostly, the connections among societies are about sharing information.

Notable examples of connections among societies include: participation in a network of Executive Directors from other societies; consulting with others when developing curricula; having MOUs with several societies to work in three focal areas (membership reciprocity, curriculum development, and access to training and licensure courses); and joint professional development workshops or seminars.

One society organized a large coalition of organizations in April of 2016—83 people from 42 different organizations to work towards the mission of producing fifty-thousand underrepresented engineers by 2025. This is a striking example of a coordinated and purposeful effort. The representative interviewee said:

> What is unique is that we are not all working together on a common program—we are saying leverage your strengths and distinctiveness and work to this common purpose.

As another example, a society benefited from an influx of foundation funding over 15 years ago to advance the educational mission of the domain. The funding allowed the society to organize and offer fellowships, early career awards, professional development for department chairs, and leadership development.

The funding also supported two education summits where people across different disciplines exchanged best practices, curricula, lab activities, and courses. These summits provided the benefit of contributions from multiple perspectives for a multidisciplinary field. Unfortunately, the program is now defunct. The society has been able to hold one summit since then and is hoping to do another.

Almost all of the societies interviewed expressed a desire to be more connected to other societies.

GAPS IN ENGINEERING EDUCATION

Most of the individuals we interviewed believe that many of the gaps in engineering education offerings they see are being addressed in some way, somewhere in the landscape. However, there were a few areas they felt need more concerted effort, such as: faculty preparation to be instructors in engineering; addressing the lack of hands-on/application experiences for undergraduates; a re-focus on the design side of engineering; an emphasis on the business side of engineering (such as financial and general business acumen); support for how to integrate new teaching technologies into the engineering classroom and for preparing engineers for new technologies; getting industry more involved in ABET; turning more attention to the two-year programs and preparedness for the workforce; support for preparing doctoral students for teaching; and preparing students in general for team-work.

One interviewee said there is generally a need for a better understanding of the 1st year experience, "*We need to change the thinking about designing for failure. We need to make student success a focus.*" A few interviewees mentioned a gap in engineering education aimed at the K-12 level. One simply said: "*We don't have the resources for k-12.*" One notable exception has a program that places high school students in labs in an effort to encourage them to pursue a technical field. They also provide resources, content, and pedagogical support to teachers at the middle school and high school levels who are teaching the subject but do not have a degree in it.

One society is very involved in filling the pre-college gap, particularly the competencies needed for high school students to be successful in college engineering. Along these lines, another interviewee voiced the need for earlier (than college) exposure to engineering and the diversity within it:

> I think what you hear often among societies at large is the feeling that there should be more at the K-12 level. We all draw from people coming out of engineering departments of Universities or Computer Science departments or Business schools, increasingly, as the industry has become more diversified. But I would say one of the challenges is that there aren't enough students coming through the pipeline who want to go through STEM programs. Both the government and other associations are trying things but they are hit or miss. It's important to give students a sense of what careers might look like in those fields.

Similarly, another interviewee expressed the need to reach younger learners if the engineering pipeline is to stay filled:

> Societies need to bring the concepts of engineering down to lower level classrooms—high school is too late. How do you do that? It requires exposure, and high school counselors are doing some of that. If kids go to [counselors] and they don't know what engineering is, or what it requires, they are turning kids off from it. The idea of exposing kids at a young age - middle school at the latest—is something that the community could be doing better. Why? To fill the pipeline.

Another interviewee noted that while they don't do any work in the pre-college realm, it is "on the list" as an area they would like to get involved in, particularly teachers.

One Executive Director lamented the current state of undergraduate engineering education:

> At the undergraduate level, are we educating the engineers of tomorrow? The curricula have been pretty stable for decades. We are not yet in the environment of tomorrow where it is about being able to learn quickly and be nimble. Are our curricula reflecting where we need to go? We need to look at the paradigm—is it adequate or does it need tweaking?

Finally, an Executive Director felt that the key missing pieces to the development of new engineers is the provision of mentors and real life education. She said:

> Students who can only solve problems from the book aren't going to go very far. They are going to run into someone that knows what they are doing. The more you get kids into that kind of [real-world] environment, the better.

FINAL NOTES

There are some interesting examples of societies that are either trying new initiatives to reach new audiences (such as a focus on early career faculty) and larger, bold initiatives that involve multiple societies and set out ambitious goals (such as the large coalition). Perhaps these societies could facilitate conversations about their experiences and stimulate thinking around innovative new programs.

It is important to keep in mind that collaborative efforts of any kind take time to both plan and get traction. They take patience and time to build trust and a collective vision. One Executive Director made this important point:

> There is never enough money and time for partnerships between like-minded societies could move the needle. It takes time. I was involved with a 10-year, multi-institutional network, and not until year 3-4 did things move. We were all doing our own thing, and then we got money to do more networking. . . . Finding partners and dancing together, that didn't happen for a while. Hosting one summit is not going to make a difference. . . . We need a long-term vision.

Appendix C

Committee and Speaker Biographies

COMMITTEE BIOGRAPHIES

Leah H. Jamieson (NAE; *chair*) is the Ransburg Distinguished Professor of Electrical and Computer Engineering at Purdue University, where she also is John A. Edwardson Dean Emerita of Engineering and holds a courtesy appointment in the School of Engineering Education. She is a member of the National Academy of Engineering and served as the 2007 president and CEO of the Institute of Electrical and Electronics Engineers (IEEE) and 2012–2016 president of the IEEE Foundation. Jamieson was a founder of the Engineering Projects in Community Service program (EPICS), a multi-university engineering design program that operates in a service-learning context. She has been recognized with the NAE's Gordon Prize for Innovation in Engineering and Technology Education, the NSF Director's Award for Distinguished Teaching Scholars, ASEE's Carlson Award for Innovation in Engineering Education, the IEEE Education Society's Harriet Rigas Outstanding Woman Engineering Educator Award, the Anita Borg Institute's Women of Vision Award for Social Impact, the IEEE Signal Processing Society's Meritorious Service Award, the Richard M. Emberson Award for contributions to IEEE's Technical Activities, and the NAMEPA Dean of Engineering Champion Award. In recognition of her leadership, the Directorship of Purdue's Women in Engineering Program was named in her honor upon completion of her term as Dean. Jamieson received a bachelor of science in mathematics from the Massachusetts Institute of Technology and a PhD from Princeton University.

Stephanie G. Adams is the 7th dean of the Batten College of Engineering and Technology at Old Dominion University. She previously served as department head and professor of engineering education at Virginia Tech and held positions at Virginia Commonwealth University and the University of Nebraska-Lincoln. Her research interests include teamwork, international collaborations, faculty development, quality control/management, and diversity in STEM. She received the American Society for Engineering Education's 2008 DuPont Minorities in Engineering Award and was invited to participate at the NAE 2006 US Frontiers of Engineering Symposium. Adams received a bachelor of science in mechanical engineering from North Carolina A&T State University, a master of engineering in systems engineering from the University of Virginia, and a PhD in interdisciplinary engineering from Texas A&M University. She is a fellow of the American Society for Engineering Education.

Marilyn Barger is the principal investigator and executive director of FLATE, the Florida Regional Center of Excellence for Advanced Technological Education, funded by the National Science Foundation and housed at Hillsborough Community College in Tampa since 2004. FLATE serves the state of Florida and is involved in outreach and recruitment of students into technical career pathways; has produced award-winning curriculum design and reform for secondary and postsecondary career and technical education programs; and provides professional development for STEM and technology secondary and postsecondary educators focused on advanced technologies. She earned a BA in chemistry at Agnes Scott College and both a BS in engineering science and a PhD in civil engineering (environmental) from the University of South Florida, where her research focused on membrane separation science and technologies for water purification. She has over 20 years of experience in developing curricula for engineering and engineering technology for elementary, middle, high school, and postsecondary institutions, including colleges of engineering. Dr. Barger serves on several national panels and advisory boards for technical programs, curriculum, and workforce initiatives, including the National Association of Manufacturers Educators' Council. She is a fellow of the American Society for Engineering Education, and a member of Tau Beta Pi and Epsilon Pi Tau honor societies. She is a charter member of both the National Academy and the University of South Florida's Academy of Inventors. Dr. Barger holds a patent and is a licensed Professional Engineer in Florida.

Steven Brown is a professor emeritus of counseling psychology at Loyola University Chicago. His research is aimed at a primary goal of promoting occupational and educational choices among diverse persons. He developed (with Drs. Robert W. Lent and Gail Hackett) Social Cognitive Career Theory to explain and predict how people develop educational and occupational interests, make educational and occupational choices, and achieve success and satisfaction in school and the workplace. Much of his current research is devoted to this theory, especially as it pertains to explaining interest, entry into, and success in STEM careers. He is also interested in international applications of vocational psychology and has worked collaboratively with scholars from Italy, Iceland, France, Switzerland, South Korea, China, and Japan to study whether measures of career indecision have the same meaning and measure the same constructs in diverse international cultures. Dr. Brown received a bachelor of arts in psychology from Muskingum College, a master of arts in experimental psychology from the University of Virginia, and a PhD in counseling psychology from the University of California, Santa Barbara.

Don P. Giddens (NAE) is dean emeritus of the College of Engineering at the Georgia Institute of Technology. He joined the Georgia Tech faculty in 1968, after two years in the aerospace industry. In 1992 he left his position as the chair of Aerospace Engineering to serve as dean of the Whiting School of Engineering and professor of mechanical engineering at Johns Hopkins University. In 1997 he returned to Georgia Tech to establish the Wallace H. Coulter Department of Biomedical Engineering, a joint department between Georgia Tech's College of Engineering and Emory University's School of Medicine. He served as the founding chair until July 2002, when he became dean of the College of Engineering. Dr. Giddens has served in a variety of professional activities involving engineering education and biomedical research. His field of research is biomedical engineering with emphasis on the cardiovascular system. He is the author of over 300 research publications, book chapters, and presentations. His professional service includes chair of the Engineering Deans Council of ASEE and president of ASEE. He chaired the NAE project that produced the report *Changing the Conversation: Messages for Improving Public Understanding of Engineering*. He received his bachelor of science, master of science, and PhD from Georgia Tech.

Asad M. Madni (NAE) was president, COO, and CTO of BEI Technologies Inc. from 1992 until his retirement in 2006. He led the development and commercialization of intelligent microsensors and systems for aerospace, military, commercial, and transportation industries, including the Extremely Slow Motion Servo Control System for Hubble Space Telescope's Star Selector System which provided the Hubble with unprecedented accuracy and stability, resulting in truly remarkable images that have enhanced understanding of the universe; and the revolutionary MEMS GyroChip® technology which is used worldwide for electronic stability control and rollover protection in passenger vehicles, saving millions of lives every year.

Prior to BEI he was with Systron Donner Corporation for 18 years in senior technical and executive positions, eventually as chair, president, and CEO. He made seminal and pioneering contributions in the development of RF and microwave systems and instrumentation which significantly enhanced the combat readiness of the US Navy (and its allies) and provided the DOD the ability (not possible with prior art) to simulate more threat-representative ECM environments for current and future advanced warfare training. Dr. Madni is an independent consultant; Distinguished Adjunct Professor and Distinguished Scientist of Electrical and Computer Engineering at UCLA; and executive managing director and CTO of Crocker Capital.

He is the recipient of major honors and awards including the IET J.J. Thomson Medal, IEEE Millennium Medal, UCSD Gordon Medal for Engineering Leadership, Mahatma Gandhi Pravasi Samman Gold Medal, IEEE AESS Pioneer Award, IEEE IMS Career Excellence Award, IEEE HKN Eminent Member and Vladimir Karapetoff Outstanding Technical Achievement Awards, and UCLA Alumnus of the Year Award. He is a member of the National Academy of Inventors, and a fellow/eminent member of 14 of the world's most prestigious academies and professional societies. He has been awarded 5 honorary doctorates and 5 honorary professorships.

He received his BS and MS degrees from the University of California, Los Angeles, PhD from California Coast University, and SE from MIT Sloan School of Management.

Thomas Perry, PE, joined ASME in 1991 as director of professional development and served as director of engineering education from 1996 until his retirement in March 2017. He oversaw ASME's work with universities and colleges of engineering and technology in the US and abroad. Working with academic, industry, and degree program accreditation volunteer leaders, Mr. Perry managed the work of the ASME Committee on Engineering

Education, its standing committees and project work groups, and served as staff liaison for the society's role in degree program accreditation through the Accreditation Board for Engineering and Technology (ABET). The work of the ASME Committee on Engineering Education engages outstanding academic and industry leaders in helping shape the quality and future of mechanical engineering education, particularly toward increasing undergraduate design-build-innovation experiences; providing greater exposure to industry practice; encouraging the employment and industry support of professors of practice/clinical professors in mechanical engineering faculties; and increasing the number of women and minority students and faculty in colleges of engineering—all part of the ASME Engineering Education Vision 2030 advocacy strategy. With over 30 years' experience in the industry, academic, and professional society communities, Mr. Perry has also served as deputy executive director for the American Society for Engineering Education (ASEE).

Anne Spence is a clinical associate professor in mechanical engineering at Baylor University. Her research program is focused on solving problems that relate to educating engineers as they navigate through the K-12 and postsecondary pipelines. She is focused on the national research agenda in engineering education which highlights engineering epistemologies, engineering learning mechanisms, engineering learning systems, engineering diversity and inclusiveness, and engineering assessment. She conducts research on and develops assessment methods, instruments, and metrics to inform engineering education practice and learning in both the K-12 and postsecondary environments. As a leader in the Project Lead the Way and FIRST communities, Dr. Spence seeks to identify best practices in educating teachers and engaging students to persevere through the STEM pipeline.

John C. Wall (NAE) has more than 35 years of industry experience in internal combustion engine technology, fuels and emissions, and global engineering organization development. Most recently, he was vice president and chief technical officer of Cummins Inc., the world's largest independent manufacturer of diesel engines and related technologies, retiring in 2015. As he progressed from research and product engineering into engineering leadership, he remained directly involved in the most critical technology programs for low emissions, powertrain efficiency and alternative fuels. He also led the growth of Cummins' technical organization from 1,000 engineers, mostly in the United States, to more than 6,000 globally, establishing new technical centers in India and China. Before joining Cummins in 1986, he led diesel and aviation fuels research for Chevron, where his team was first to discover the important contribution of fuel sulfur to diesel particulate emissions. Today he stays active technically as an advisor for the DOE Joint BioEnergy Institute and Co-Optima Program, the Cyclotron Road energy incubator at Lawrence Berkeley Laboratory, in the work of the National Academies, and as chair of the Cummins Science and Technology Council. He has been recognized for his technical contributions by election to the NAE and as a fellow of the Society of Automotive Engineers. He has received the SAE Horning Memorial Award and Arch T. Colwell Merit Award for research in the area of diesel fuel effects on emissions, and Franz F. Pischinger Award for Powertrain Innovation, the ASME Soichiro Honda Medal for significant engineering contributions in the field of personal transportation, and the California Air Resources Board Haagen-Smit Clean Air Award and US EPA Thomas W. Zosel Individual Achievement Award for career accomplishments in diesel emission control. Dr. Wall studied mechanical engineering at the Massachusetts Institute of Technology, where he received his SB and SM degrees in 1975 and ScD in 1978.

Gregory N. Washington is the Stacey Nicholas Dean of the Henry Samueli School of Engineering at the University of California, Irvine. Prior to his arrival at UC Irvine, he served as the interim dean for the College of Engineering at the Ohio State University (OSU), where he provided visionary leadership, oversight, and management for one of the nation's largest and highest-ranked engineering programs. He joined the faculty at OSU in 1995, became the associate dean for research in 2005 and was appointed interim dean in 2008. As a professor of mechanical engineering, his research is in the design and control of smart material systems, the design and control of hybrid electric vehicles and the design of smart electromagnetic systems. Washington has been involved in multidomain research for the last 20 years. His core area of interest is dynamic systems, with an emphasis on modeling and control of smart material systems and devices. He has been involved in the design and control of mechanically actuated antennas, advanced automotive systems incorporating smart materials, hybrid electric vehicles, and structural position and vibration control with smart materials. Dr. Washington received a bachelor of science, master of science, and PhD in mechanical engineering from North Carolina State University.

SPEAKER BIOGRAPHIES

Barbara Bogue is cofounder and director of the AWE Project and retired associate professor of engineering science and mechanics and women in engineering at Penn State University Park. Previously, as director of the Penn State Women in Engineering Program, Ms. Bogue led the establishment of recruitment, retention, and development activities for girls and women. Her work was recognized with several awards, most notably the Presidential Award for Excellence in Science, Mathematics, and Engineering Mentoring. Ms. Bogue develops and presents workshops on assessment and faculty development and evaluates and consults on interventions that aim to increase the participation and progression of women and underrepresented minorities in STEM. She publishes frequently on topics related to engineering education, equity, and assessment and serves on several national advisory boards. She holds an MSc in social sciences from the University of Southampton. Ms. Bogue lives with her husband in Alexandria, Virginia, and enjoys writing, gardening, and developing engineering activities for her six grandchildren.

Elliot Douglas is the NSF program director for Engineering Education. He is also associate professor of environmental engineering sciences and Distinguished Teaching Fellow at the University of Florida. He is director of the Engineering Education Collaborative, which brings together faculty interested in all aspects of engineering education, from improving their teaching to conducting education research. His research interests lie at the intersection of education research and engineering education practice. His work aims to understand complex thinking processes and learning in students, and to use this information to design effective teaching practices, and includes research in critical thinking, active learning, and problem solving. He has recently begun a project to examine the culture of inclusion in high-tech firms through the narratives of minority engineers. He also conducts work on qualitative methodologies in engineering education research. He has published a textbook, *Introduction to Materials Science and Engineering: A Guided Inquiry*, which provides faculty teaching Introduction to Materials a means to easily incorporate active learning techniques into their classrooms. He has been involved in faculty development activities since 1998, most recently presenting workshops on active learning through the POGIL Project. Dr. Douglas received SBs in materials science and engineering and MSE and Music from MIT in 1988, and his PhD in polymer science and engineering from UMass-Amherst in 1992. He then worked at Los Alamos National Laboratory for four years before joining the University of Florida in 1996. He has served as deputy editor of the *Journal of Engineering Education* and chair of the Educational Research and Methods Division of ASEE.

Darryll J. Pines is dean and Nariman Farvardin Professor of Aerospace Engineering at the Clark School of Engineering at the University of Maryland. He arrived at the Clark School in 1995 as an assistant professor and in 2006 became chair of the Department of Aerospace Engineering. As dean since 2009, Dr. Pines has led the development of the school's strategic plan and improved teaching in fundamental undergraduate courses, raised student retention, achieved success in national and international student competitions, placed new emphasis on sustainability engineering and service learning, promoted STEM education among high school students, increased the impact of research programs, and expanded philanthropic contributions to the school. Today, the school's one-year undergraduate retention rate and six-year graduation rate are 87.6 percent and 74.2 percent respectively; the university's Solar Decathlon team placed first worldwide in a recent competition against other leading universities; the university's Engineers Without Borders chapter is considered one of the nation's best; and the Engineering Sustainability Workshop launched by Pines has become a key campus event. At the national level Pines has testified before Congress on STEM education and has led an effort as part of the American Society for Engineering Education Deans Council's K-12 STEM Committee to develop a potential College Board AP exam in engineering. He is secretary of the executive committee of the National GEM Consortium, a national nonprofit providing programming and full fellowships to support increasing untapped domestic human capital at the graduate level in STEM fields. Dr. Pines' current research focuses on structural dynamics, including structural health monitoring and prognosis, smart sensors, and adaptive, morphing, and biologically inspired structures, as well as the guidance, navigation, and control of uninhabited aerospace vehicles. He is a fellow of the Institute of Physics, American Society of Mechanical Engineers, and American Institute of Aeronautics and Astronautics, and has received an NSF CAREER Award. He received a BS in mechanical engineering from the University of California, Berkeley, and MS and PhD degrees in mechanical engineering from the Massachusetts Institute of Technology.

Betty Shanahan is a consultant to the executive vice president of administrative services at Michigan State University. From 2002 to 2014, she was executive director and CEO for the Society of Women Engineers (SWE). Prior to SWE, she spent 24 years in executive management, engineering management, development, and marketing in the electronics and software industries. In 2010 she was the first woman to receive the Claud R. Erickson Distinguished Alumnus Award from the College of Engineering at Michigan State University. In 2013 she received an honorary doctor of science from the University of Connecticut. In 2016 the American Association of Engineering Societies recognized her contributions to the engineering profession and diversity in engineering with its Chair's Award. Ms. Shanahan earned a BS in electrical engineering from Michigan State University, a master of software engineering from the Wang Institute of Graduate Studies, and an MBA in strategic management from the University of Chicago Booth School of Business.

Appendix D

Participants List

Stephanie Adams*
Old Dominion University

Colleen Agan
Institute of Transportation Engineers

Ashok Agrawal
American Society for Engineering Education

Lucy Alexander
American Institute of Chemical Engineers

Karin Anderson
Society for the Advancement of Material and Process Engineering, North America

Thomas Attard
University of Alabama at Birmingham

Nancy Bakowski
American Chemical Society

Matt Barrios
Tulane University

Elizabeth Bierman
Society of Women Engineers

Barbara Bogue
Society of Women Engineers – Advancing Women in Engineering

Virginia Booth-Womack
National Association of Multicultural Engineering Program Advocates

* Organizing Committee members and NAE project staff

Stephen Brock
American Institute of Aeronautics and Astronautics

Nichol Campana
ASM International

W. Bernard Carlson
University of Virginia

Jenna Carpenter
Campbell University

Chris Ciuca
SAE International

Wendy Cowan
American Association of Engineering Societies

Courtney Day
Institute of Transportation Engineers

Burton Dicht
Institute of Electrical and Electronics Engineers

Elliot Douglas
National Science Foundation

Arash Esmaili Zaghi
University of Connecticut

Robert Fine
American Nuclear Society

Marc Fry
Granta Design Ltd.

Don Giddens*
Georgia Institute of Technology

Mark Golden
National Society of Professional Engineers

Donald Hayes
University of Nevada, Las Vegas

Robin Hensel
West Virginia University

Charles Hickman
Accreditation Board for Engineering and Technology

James Hill
Iowa State University

Abby Ilumoka
National Science Foundation

Leah Jamieson*
Purdue University

Kenan Jarboe*
National Academy of Engineering

Libby Jones
Engineers Without Borders USA

Burk Kalweit
American Academy of Environmental Engineers and Scientists

Linda Katehi
University of California, Davis

Shankar Krishnan
International Federation of Medical and Biological Engineering

Glenda La Rue
Women in Engineering ProActive Network

Lawrence Larson
Texas State University

Aisha Lawrey
American Society of Mechanical Engineers

Catherine Leslie
Engineers Without Borders USA

Calvin Li
Villanova University

Alvin Lomibao
University of Rochester

Anastasios Lyrintzis
Embry-Riddle Aeronautical University

Asad Madni*
BEI Technologies Inc.

Bill Mahoney
ASM International

Albert Manero
National Academy of Engineering

Kevin McLaughlin
University of Connecticut

Angela Moran
United States Naval Academy

Donna Nelson
American Chemical Society

Harriet Nembhard
Oregon State University

Leslie Nolen
American Society of Civil Engineers

Jim O'Brien
American Society of Civil Engineers

Steve Olson
Science writer

Karen Panetta
Tufts University/IEEE

Thomas Perry*
American Society of Mechanical Engineers

Karen Peterman
Karen Peterman Consulting Co.

Bonnie Peterson
SPIE – The International Society for Optical Engineering

Olga Pierrakos
National Science Foundation

Darryll Pines
University of Maryland, College Park

Melissa Prelewicz
American Association of Engineering Societies

Jane Prey
Association for Computing Machinery

Rama Ramakrishna
National Academy of Engineering

Teri Reed
Women in Engineering ProActive Network

Karl Reid
National Society of Black Engineers

Proctor Reid*
National Academy of Engineering

Charles Reinholtz
Embry-Riddle Aeronautical University

Randi Rosenbluth
Society of Women Engineers

Mehran Sahami
Association for Computing Machinery

Betty Shanahan
Society of Women Engineers – Advancing Women in Engineering

Amelia Shaw
National Academy of Engineering

David Silverstein
University of Kentucky

Anne Spence*
University of Maryland, Baltimore County

David Spencer
wTe Corporation

Heidi Steinhauer
Embry-Riddle Aeronautical University

Constance Thompson
National Society of Black Engineers

Massood Towhidnejad
Embry-Riddle Aeronautical University

Victoria Valentine
Society of Fire Protection Engineers

Chester Van Tyne
Colorado School of Mines

Kodi Verhalen
National Society of Professional Engineers

John Wall*
Cummins Inc.

Kristine Ward
SME

Gregory Washington*
UCI Samueli School of Engineering

William Wepfer
Georgia Tech/American Society of Mechanical Engineers

Phillip Westmoreland
North Carolina State University

Giovanni Zangari
University of Virginia

Tasha Zephirin
North American Marine Environment Protection Association